CW01501951

Preface

Reflections on Nature
in a Human World

This is a book about the natural and social history of the River Tees in the North-East of England, but it's neither a provincial natural history nor a local history book – there are a great many people more qualified than me to write either of those books. That said, in making a journey along the Tees, from source to mouth, we'll discover a wealth of local social and natural history, but these stories connect us to bigger national and global stories, which in turn allow us to reflect on how the natural and human worlds have intersected throughout history to create the world we inhabit today. Many stories from the Tees Valley are of great relevance in this wider context – far more than I could ever have hoped for when I began this project. The Tees, it turns out, is the perfect place for such a venture. This tale of nature and humanity along the Tees is a story of the whole of Britain and, at times, of the whole planet.

I've spent the bulk of my career – some forty years as I write this – making wildlife and science documentary films around the globe. I've worked in the heart of the Amazon and on the dizzying heights of the Andes, in the baking deserts of Australia, Africa and America, and in the freezing Arctic in the depths of winter. I've visited tiny, remote islands in the vastness of the Atlantic and Pacific Oceans and explored the rainforests of South-East Asia. These places, along with the local people with whom I worked, have shaped my views on the state of the natural world and on our responsibilities as citizens of Planet Earth. So, why have I chosen to wrestle with these large ideas in a book about one small, relatively unknown river in an unfashionable corner of the country?

I was born in Middlesbrough, on the southern bank of the lower Tees, and lived there until I was eighteen years old. Although I haven't lived in the North-East for the last fifty years, my whole family remained there – so, despite living in Bristol in the South-West since leaving Middlesbrough, I've made the more-than-500-mile round trip many times a year for five decades. The result is that, if people ask where I'm from, I still reply 'Middlesbrough'. To paraphrase an old expression, you can take the lad out of Middlesbrough, but you can't take Middlesbrough out of the lad. The North-East remains a special place for me – it always feels like coming home. So, this book is also a strongly personal one. The Tees Valley is where my life-long love of nature was born and nurtured, which makes it the most appropriate place for me to share my views on nature and humanity, drawn from experiences around the world in the half-century since I left the North-East.

Social and natural histories are intertwined in the field of environmental history. I've always had a deep interest in this topic because I believe it has much to teach us about living in the modern world. One of my earlier books, *Paradise Found: Nature in America at the Time of Discovery,* for the University of Chicago Press, paints a picture of the natural world across North and Central America from the arrival of European settlers in the late fifteenth century to the present day. It documents an extraordinary and often unsuspected abundance of nature in the past and provides a stark illustration of just how much we've diminished nature in our modern world – more than even the most pessimistic of us might imagine.

My journey along the Tees uncovers a parallel but very different story from that told in *Paradise Found.* The estuary of the

STEEL RIVER

STEVE NICHOLLS is an award-winning nature
documentary producer and director and is a
Fellow of the Royal Entomological Society of
London. He is the author of *Paradise Found:
Nature in America at the Time of Discovery*,
*Flowers of the Field: A Secret History of Meadow,
Moor and Woodland* and *Planet Insect: How
Insects Conquered the Earth.*

Also by Steve Nicholls

*Paradise Found: Nature in America at the
Time of Discovery*

*Flowers of the Field: A Secret History of Meadow, Moor
and Woodland*

Planet Insect: How Insects Conquered the Earth

STEEL RIVER

Walking the Tees:
A Journey Through Nature
in a Human World

STEVE NICHOLLS

HEAD
ZEUS

An Apollo Book

First published in the UK in 2025 by Head of Zeus Ltd,
part of Bloomsbury Publishing Plc

Copyright © Steve Nicholls, 2025

The moral right of Steve Nicholls to be identified as the author
of this work has been asserted in accordance with the Copyright,
Designs and Patents Act of 1988.

All rights reserved. No part of this publication may be: i) reproduced or transmitted
n any form, electronic or mechanical, including photocopying, recording or by means
of any information storage or retrieval system without prior permission in writing from the
publishers; or ii) used or reproduced in any way for the training, development or operation
of artificial intelligence (AI) technologies, including generative AI technologies. The rights
holders expressly reserve this publication from the text and data mining exception
as per Article 4(3) of the Digital Single Market Directive (EU) 2019/790.

9 7 5 3 1 2 4 6 8

A catalogue record for this book is available from the British Library.

ISBN (HB): 9781804542613
ISBN (E): 9781804542590

Photographs © Steve Nicholls
Illustrations by Victoria Coules
Typeset by DivAddict Solutions Ltd

Printed and bound in Great Britain by
CPI Group (UK) Ltd, Croydon CR0 4YY

MIX
Paper | Supporting
responsible forestry
FSC® C013604

Bloomsbury Publishing Plc
50 Bedford Square, London, WC1B 3DP, UK
Bloomsbury Publishing Ireland Limited,
29 Earlsfort Terrace, Dublin 2, D02 AY28, Ireland

HEAD OF ZEUS LTD
5–8 Hardwick Street
London EC1R 4RG

To find out more about our authors and books
visit www.headofzeus.com
For product safety related questions contact productsafety@bloomsbury.com

STEEL RIVER

Contents

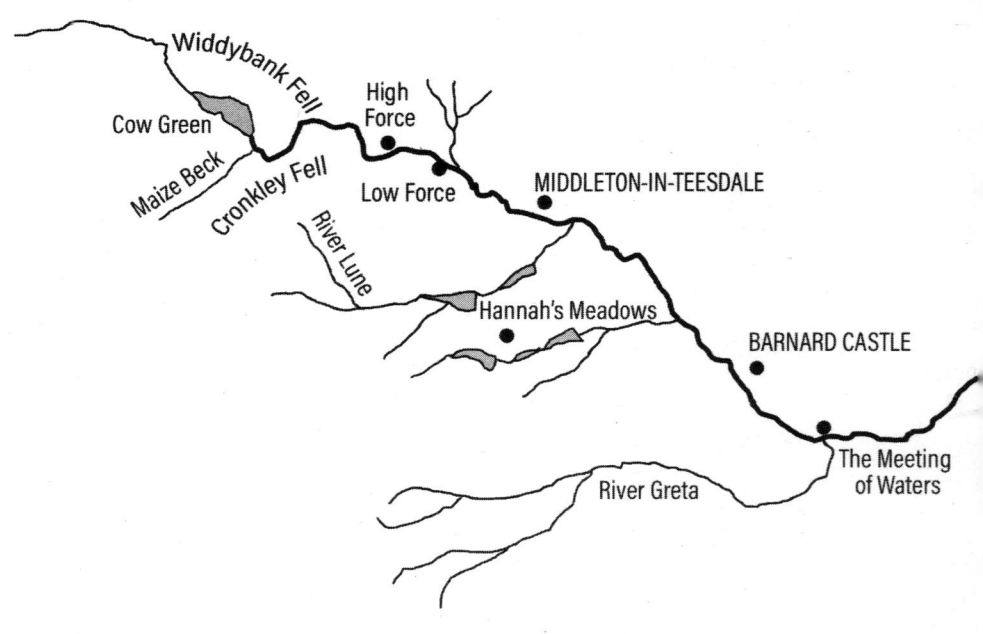

Widdybank Fell

Cow Green

Maize Beck

Cronkley Fell

High
Force

Low Force

River Lune

MIDDLETON-IN-TEESDALE

Hannah's Meadows

BARNARD CASTLE

The Meeting
of Waters

River Greta

10 km

River Tees

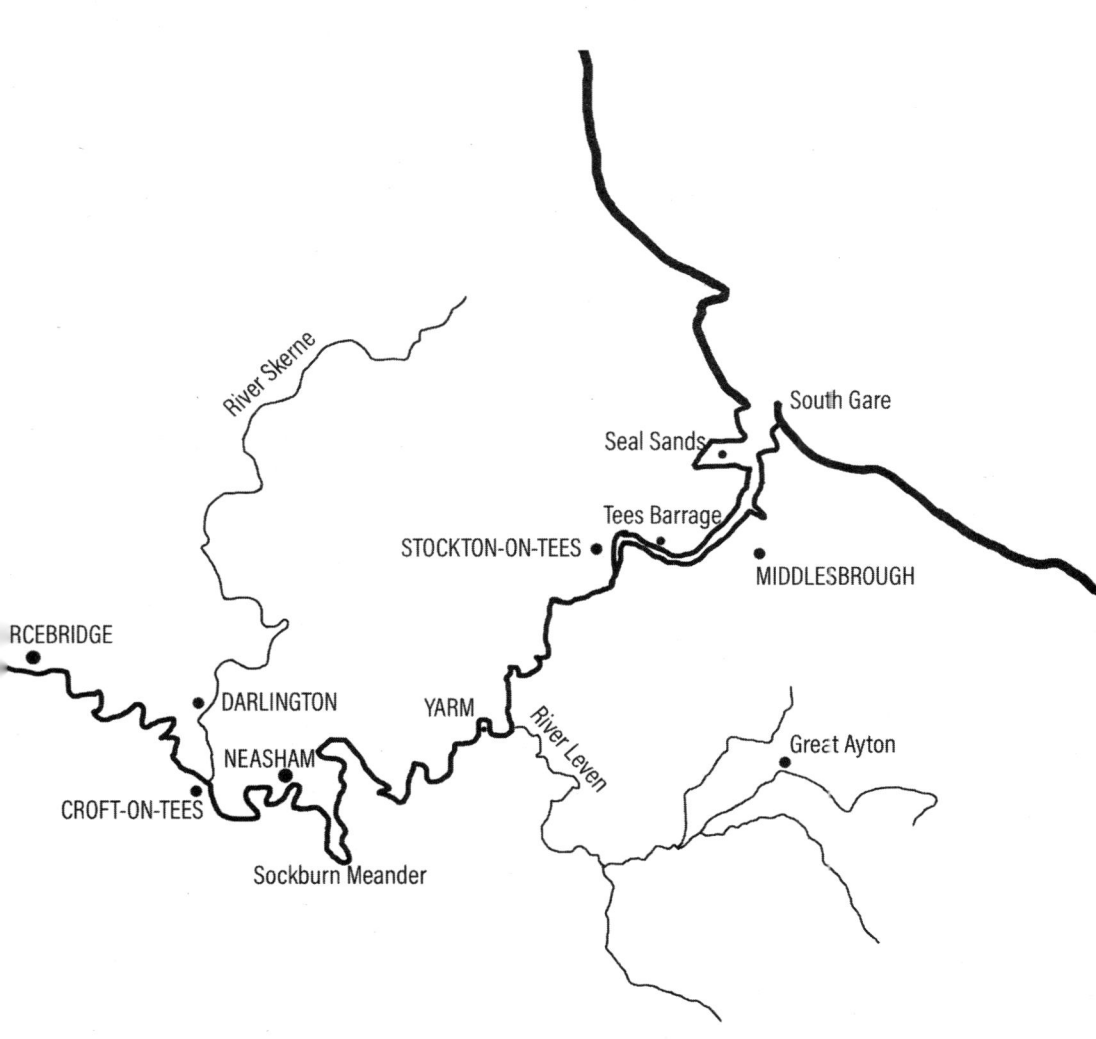

River Skerne

South Gare

Seal Sands

Tees Barrage

STOCKTON-ON-TEES

MIDDLESBROUGH

RCEBRIDGE

DARLINGTON

YARM

River Leven

Great Ayton

NEASHAM

CROFT-ON-TEES

Sockburn Meander

Tees has seen a colossal human impact, as great as anywhere in the country, as heavy industry has spread along both banks. In stark contrast, just eighty-five miles away, the Tees rises in the wild country of the high Pennines, although, as we'll discover, even in these remote fells, wildlife has been dramatically reduced. However, since I'm a born optimist, this story is certainly not relentlessly gloomy.

We'll discover many reasons for hope, although it's not enough just to hope. We all have an active role to play in restoring our world. Ultimately our journey leads to the inevitable conclusion that we all – each and every one of us – need to see the world and our place in it in a very different way. This new worldview will need to be so radically different from our current perspective that it might seem like an impossible dream. Yet, as unattainable as it may look, a growing number of people in many disciplines have already set out on this journey. We'll meet many of them on our own travels along the Tees. It turns out we know a great many of the answers; we just have to put them into practice.

The Tees is an excellent place to appreciate all these elements – the good, the bad and the ugly. Despite being buried under industry and suffering some of the worst pollution in the country, nature has hung on along the lower Tees. In the twenty-first century, with an enlightened and groundbreaking partnership between conservation organisations and individual industries, the Tees Estuary once again boasts some spectacular wildlife sites. Given even half a chance, nature is often quick to rebound – a fact which has encouraged me to continue fighting to create a meaningful place for nature in what today is predominantly a human world.

I was a trustee of the Avon Wildlife Trust for eleven years, eight of them as its vice-chair, a time in which I realised just how little we're doing to preserve and nurture the life support systems of our planet. That's not an indictment of the wildlife trusts or the other NGOs, such as the RSPB, Plantlife,

Buglife, the Woodland Trust and many others, without whom we'd be incalculably worse-off. All these organisations punch well above their weight, but they operate in a world where, despite a rapidly approaching point of no return, environmental issues are nowhere near high enough on public or government agendas.

What can you expect in the following pages? Well... quite a lot. We'll be mixing deep geology with politics, natural history with economics, social history with science, and religion with industrial history. It might sound like these subjects are unrelated, but the journey we're about to make will show us that if we're to repair the damage we've done to our only home, all these aspects of our lives must become part of one joined-up worldview. If we're to survive the crises that we now face, we need to fundamentally reassess each of these topics. Our new perspective on the world must incorporate entirely new ways of thinking about the environment, economics, politics and social justice and link all these disparate aspects of our lives into a single coherent philosophy.

So how does this patchwork of ideas fit together and what does this new mindset look like – and what does the Tees have to do with any of this? Let's take a walk...

Part I

ESTUARY

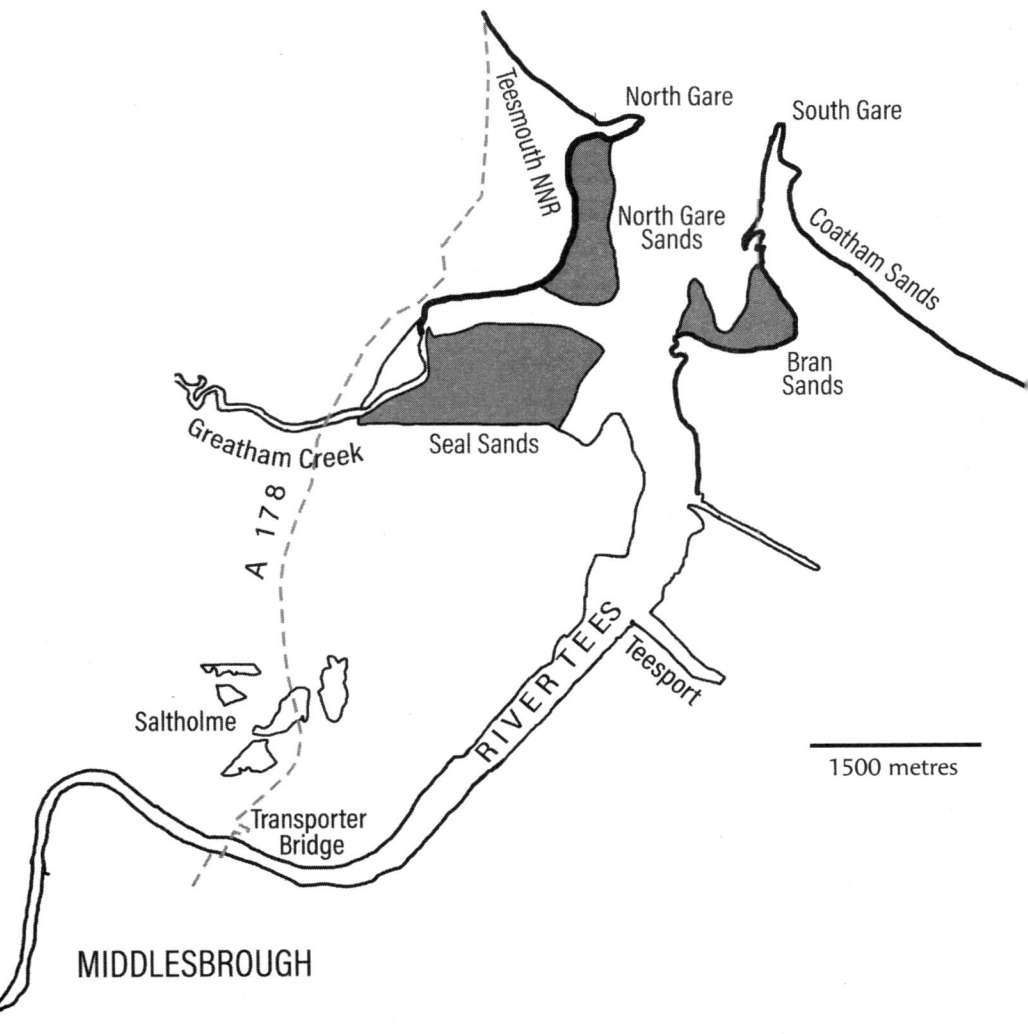

I

From Senility to Salinity

The Tees Meets the North Sea –
Bridges, Birds and the Birth of a Naturalist

The views that I cherish, those stark purple moors
A stranger's eye could nowt but please
And a contrasting beauty is free to behold
Where those hills meet the Valley of Tees

VIN GARBUTT,
'The Valley of Tees', 1972

A biting wind howls across mudflats and grazing marsh, slicing with ease through the latest 'windproof' technology that lured me into buying this expensive jacket. Even an equally expensive base layer of merino wool does little to keep me warm. This is a Teesmouth wind, a nor'-easter, blowing straight from the Arctic, bringing a taste of the far north to this corner of North-East England. It's December and under leaden skies it barely seems to get light at all. In any case, there's little colour out on the estuary at this time of year – a scene in shades of grey, punctuated only by the occasional orange flame from a flare stack, startlingly bright on this gloomy day.

This is no scene of wild beauty. To the north I can make out the square block of Hartlepool's nuclear power station. All around me, fields and marshes are backed by a tangle of pipework and chimney stacks. Scraggy-looking ponies hunker down behind wind-blown shrubs and, behind them, to the south of where I'm standing, is the unmistakable silhouette of the Transporter Bridge, spanning the Tees between Port Clarence and Middlesbrough. This is (or was) one of the largest areas of heavy industry in the whole of Europe and it's certainly not pretty, yet for me this is a special place. I was born on the other side of the Transporter from where I now stand, in Middlesbrough, and spent the first eighteen years of my life there. As unlikely as it might seem, these vast industrial landscapes sparked my interest in the natural world, an interest that grew to be a career, first in academia then, for the last forty years, as a wildlife film-maker, photographer and writer.

Nor am I the only soul braving the vicious wind. Walking back along a rough track from a hide that overlooks what remains of the mudflats of Seal Sands after big industry has reclaimed most of it, I meet a couple of other birdwatchers, who look as frozen as I feel.

'Alright. Anything about?' they enquire.

So far, I've only seen what I'd expect – flocks of dunlin and knot in flight, grey birds against a grey sky, twisting and turning in unison as they settle on the mudbanks. Knot. A curious name, its derivation is the same as the birds' Latin name: *Calidris canutus*. Yet unlike King Canute, they don't try to stop the tide. Instead, they make good use of it, following it as it flows out across the mudflats, probing newly exposed mud for invertebrates. And there are shelducks. There are always shelducks, a distant line of black-and-white dots, far out in the estuary, although through the scope each resolves into its elegant uniform of black, white and chestnut.

'There's a black-necked grebe on Saltholme Pool – saw it there about an hour ago – should still be around', they tell me.

I'm grateful for the information. Teesmouth is scrutinised by a small army of experienced and enthusiastic birdwatchers and little of interest goes unnoticed. Back in 1960, the Teesmouth Bird Club was originally founded to record any unusual finds on or around the estuary. Even when I first began birdwatch-ing here in the late 1960s, the whereabouts of any rarity was quickly passed around between the club's members and beyond. Today, with the internet and texting, the transmission of such information is instantaneous. A black-necked grebe isn't particularly unusual (not a 'mega' in twitcher parlance), and it won't look much in its winter plumage – more shades of grey – although it's still well worth the time to pay it my respects.

However, the Tees Estuary has seen more than its fair share of real rarities, waifs and strays blown off-course by conditions like those out here today. And some of these were well and truly off-course. In August 1982, a long-toed stint paid a visit – as far as is known, only the second one ever to set its long-toed feet on British soil.* These birds breed in Siberia and normally migrate to Australasia in winter, so this little lost scrap of feathers drew flocks of birdwatchers from all over the country. To date, there are still only three or four records for this species anywhere in Britain. In 1989, a double-crested cormorant from North America splashed down on Teesside, this time a first for the whole of the UK, which caused an even bigger stir. I'm told that two birders flew over from Finland especially to see this rare transatlantic visitor – twice, since they failed to see it on their first trip. The list of Teesmouth rarities includes a great many other exotic species – Icterine warblers, rose-coloured starlings, rollers, subalpine warblers and more – plenty to keep the Teesmouth Bird Club busy over the years.

Many of the smaller migrants turn up on the other side of the river, along the southern breakwater at the river-mouth, known as South Gare. There's a patch of scrub here – the Shrike Bushes – named for migrating red-backed shrikes that occasionally use them as a temporary shelter before heading on to their final destination. Many of the bushes were originally planted by bird ringers and now provide a similar refuge for all kinds of other migrants, from those regulars that pass through this area every year to those that have wandered far from home and cause a real stir. I once saw a cuckoo from North America, a black-billed cuckoo, in a park near South Gare, and although I don't consider myself a twitcher, seeing such birds that have survived apparently impossible journeys always brings home how tough these seemingly fragile creatures really are.

In the welcome shelter of a hide that overlooks Saltholme Pool, now part of a well-developed RSPB reserve, I spot the black-necked grebe. Actually, it doesn't take much birdwatching prowess. A dozen scopes are pointed at it when I arrive. Even so, I'm pleased to see it – not the first I've seen on the

---

* For a long time the Tees record was seen as a national first, until an older record from Cornwall, previously rejected, was accepted by the committee that oversees such things.

estuary and hopefully not the last, but it starts me thinking about how a boy from a council estate in Middlesbrough got so interested in nature.

Middlesbrough might seem an unlikely place to nurture a love of all things wild – mind you, it does get pretty wild most Saturday nights. It's also the butt of endless jokes about its poor quality of life – and I'm pretty sure I've heard all of them over the years. In the North-East, we're the poor relations to Newcastle, thirty miles to the north. Outside the North-East, most people have no idea where Middlesbrough is. Anybody who has heard of it has probably done so in the context of it being one of the least desirable places in Britain to live. People from Middlesbrough are known as 'Smoggies', a nickname from the heyday of big industry along the Tees, when chemical works and steelworks belched out all manner of nasty fumes. And to be fair, after I left Middlesbrough in the early 1970s, I could definitely smell the air when I came back to visit my family. An acrid odour, so strong it made the air feel thick, always greeted me as I left the train at Middlesbrough station, although after a few hours, I got used to it again and stopped noticing.

The 1960s and 1970s were bad times for pollution of land, air and water on Teesside, which meant that as I was growing up, nature was in retreat. Yet most people here depended on these same industries for a living, however hard and meagre it may have been. The demise of many of the industries along the Tees over the last four decades has spelled real hardship for people here, although at least today the air smells cleaner* and the river runs cleaner too, even if not yet nearly clean enough. This apparent conflict between vibrant ecology and social wellbeing makes the River Tees the perfect place to reflect on

---

* Detailed monitoring of key atmospheric pollutants over the last two decades backs up my impressions. These pollutants now generally fall within national air quality standards. R. Woods. 'The State of the Natural Environment in the Tees Estuary'. INCA. 2011.

nature's place in a human world as well as humanity's place in the natural world. Any recipe for a viable future must reconcile these two aspects of life and not set one against the other. In any case, as we'll discover later, social justice and environmental health are two sides of the same coin.

Middlesbrough might not make it onto any lists of the best places in Britain to live, but the antidote to such attitudes is the self-deprecating sense of humour so typical of Smoggies. From the hide at Saltholme, I can see the outlines of the Riverside Stadium, home to Middlesbrough Football Club – a team I've supported all my life, through thick and thin, for better or worse, although in retrospect it seems mostly worse. When I lived here, Boro played at Ayresome Park in the centre of town. Today, their new stadium is just the other side of the river from where I'm sitting – close enough for me to hear the crowd roar on those occasions when birdwatching seems like a less stressful way to spend a Saturday afternoon. In 2004, an unusually successful season, I went to watch Boro playing in the League Cup final at the Millennium Stadium in Cardiff,* just across the River Severn from Bristol, my home for the last fifty years. I couldn't help but smile when I saw, held high above the Boro fans packed into the stadium behind me, banners reading 'Smoggies on Tour – We're only here for the fresh air'. And we won!

By the time I moved from Middlesbrough in 1973, at the age of eighteen, to study zoology at the University of Bristol, I was already deeply interested in many aspects of the natural world. This was thanks in no small part to one man, who happened to be a founder member of the Teesmouth Bird Club. Ken Smith is now something of a legend in natural history circles on Teesside, but when I first met him, he worked at Imperial Chemical Industries (ICI), one of the biggest of the industries

---

* Wembley Stadium, the usual home for such prestigious matches, was being rebuilt at the time.

which once lined both banks of the Tees. He joined ICI as a chemist but soon found a more fitting role as its ecological advisor, eventually playing a key role in founding a ground-breaking conservation organisation, the Industry Nature Conservation Association (INCA).* INCA is a partnership between Teesmouth's industries and conservation organisations that has helped mitigate the effects of industry on nature on the Tees Estuary over the past few decades. In my youth, however, Ken ran evening classes on birdwatching in Middlesbrough's Dorman Museum. Although these were really aimed at adults, I enrolled. I think I was only about fourteen at the time and felt a little daunted, but was quickly taken under the wing, to use an obvious metaphor, of several of the other class members.

Ken's evening classes were held in the Nelson Room, where we were surrounded by a collection of stuffed British birds. Specimens of each species were mounted in separate display cases, often on a recreated piece of their natural habitat. Some even had painted backdrops, turning them into windows onto places and birds I longed to see for real. These cases lined all four walls of the Nelson Room and beneath them, running right round the room, were glass-topped cabinets displaying a dazzling variety of birds' eggs. I was instantly drawn to the strange yet beautiful eggs of guillemots, each and every one with a different pattern. I recently returned to the Dorman Museum after its extensive refurbishment, and was heartened to discover that the Nelson Room was still intact, preserved in all its Edwardian splendour. The guillemot eggs were still there, as was that aroma of polished parquet flooring that I remembered from my childhood.

The Nelson Room is named after Thomas Hudson Nelson, a collector of birds and eggs in the late nineteenth and early

---

* INCA was formed in 1988, when ICI and the Nature Conservancy Council (now Natural England) began to work together. Today, many businesses and conservation organisations, including the Teesmouth Bird Club and the RSPB, are part of a much-expanded partnership.

twentieth centuries. He was also the author of *The Birds of Yorkshire*, a very useful source of information on how the birds of the Tees Valley were faring at the very start of the twentieth century. He'd moved from Bishop Auckland in County Durham, on doctor's advice, to Redcar, a small town on the coast close to where the Tees empties into the North Sea, and he soon became well known in the area. He bequeathed his extensive collection to Dorman Museum in 1914. Part of this was his large collection of guillemot eggs, a few of which were in the displays that so entranced me as a young teenager. In the nineteenth century, guillemot eggs, which are more variable in pattern and colour than other birds' eggs, were collected with zealous passion to line the cabinets of zoologists. Most of Nelson's came from Bempton Cliffs – further south on the Yorkshire coast and today another RSPB reserve – where guillemots nest on narrow ledges. In Nelson's day, gangs of men descended these imposing cliffs on ropes, protected from rock falls only by cloth caps stuffed with straw, to collect guillemot eggs (which are three times the size of hen's eggs) to sell as food. But any that were unusually marked were put to one side for Nelson or other egg enthusiasts, who would pay far more for such prizes. Such was the passion for guillemot eggs that on one occasion a collector, waiting on the clifftop for the 'climmers' (as the egg collectors were called) to ascend, pulled his gun on rival collectors to ensure he had the pick of the bunch.

In the Nelson Room, I could see why these beautiful and curious eggs were so eagerly sought by Victorian collectors. They're covered in an endless variety of blotches and scribbles, usually etched onto a blue-green or buff background. The climmers discovered that each guillemot laid eggs with her own unique markings year after year. In some cases, they collected eggs from the same females for twenty years – so here was an early example of citizen science, although it meant that those unfortunate females never managed to rear a single chick in all that time. It's now thought that these distinctive individual patterns help each bird recognise its own egg on the crowded and bustling ledges.

Guillemot eggs are also a very unusual shape, sharply pointed at one end, making them distinctly pear-shaped. When I began my studies in the Nelson Room, Ken told us that this was to allow the egg to roll in a tight arc, so it wouldn't roll off the narrow nesting ledge and plunge into the sea a few hundred feet below. This reasoning made sense to me, and so the story remained until 2016, when a Sheffield University ornithologist would debunk this long-held idea, which had its roots as far back as the early nineteenth century. In 1831, egg collector William Hewitson had discovered that if he set his guillemot eggs spinning, they would rotate in place, allowing the egg to remain safely on its narrow ledge. However, had he bothered to try the same experiment with any of his other eggs, he would have found exactly the same thing. All his eggs had been blown, and so they were empty shells, which behaved nothing like fresh eggs. Even so, his idea caught on and was widely accepted for the best part of a century, until, in the 1940s, Russian biologist Lew Belopol'skii refined this thinking. He demonstrated that a fresh guillemot's egg, given a small push, rolled in a tight arc and that this must be the reason why guillemot eggs remain safely perched on their precarious nesting ledges. Although it's true that guillemot eggs do roll in a tighter arc than those of the closely related razorbill (which doesn't usually nest on such narrow cliff ledges), this observation only holds up if the eggs are rolled on unnaturally flat surfaces. On a rough and uneven rock ledge, rolling eggs of either species behave much more unpredictably. Nor is it true that guillemot eggs are entirely safe from rolling off narrow ledges.

In times past, steamers carrying tourists to the base of teeming nesting cliffs would blast their horns or even fire off shotguns to provide the onlookers with the spectacle of thousands of birds erupting from the cliff face. Unfortunately, as the panicked birds took flight, there was also a spectacular (and disastrous) cascade of eggs from the nesting ledges. Clearly the widely held theory accounting for the guillemot's pear-shaped egg needed revisiting and ornithologist Tim Birkhead came up with several plausible alternative ideas.

First, he suggested that their unusual shape may help protect the egg from impacts. Guillemots are not the most agile of fliers. Their short, stubby wings have been shaped by a guillemot's need to 'fly' underwater in pursuit of small fish. But in air, this wing shape means that they have no option but to fly fast and straight. So, landing on a narrow ledge on a sheer cliff is fraught with difficulty. Often, an incoming guillemot simply crash-lands on the ledge, or onto any guillemot that might already be there, incubating its egg; perhaps the shape of the egg helps it withstand the force of such collisions. However, it's not clear how their extreme pear shape helps to strengthen the egg and, anyway, there are less extravagant ways of bomb-proofing an egg. The easiest way is simply to lay down a thicker shell, which is exactly what guillemots do – their eggs have thicker shells than those of razorbills. Guillemot eggs obviously do need to stand up to frequent impacts, although it doesn't seem that their unusual shape is much help.

On the other hand, the extreme pear shape causes a guillemot's egg to sit with the blunt end raised off the ledge, where filth and muck accumulate. The large blunt end may therefore remain cleaner than the pointed end. The shell is permeated by tiny pores which allow the growing embryo to breathe, but they're much more densely packed in the blunt end, so preventing these vital pores from becoming clogged is obviously a good idea. However, yet again, observations don't bear out this ingenious suggestion; as the season progresses, the whole egg becomes coated in muck.

While handling the fresh eggs of both guillemots and razorbills, Birkhead was struck by a revelation. The weight distribution and shape of a guillemot's egg means that more of the shell lies in contact with the ground than would be the case for a more conventionally shaped egg. This makes a guillemot egg more stable, particularly on sloping ledges. A razorbill's egg simply slides off a sloping surface, whereas a guillemot's egg, with greater ground contact, stays put. Razorbills often tuck themselves away in more secure crevices, leaving guillemots to crowd onto narrow, exposed ledges. Colonies of guillemots can be enormous, and, in such cases, nesting ledges are at a premium but, thanks to their unique eggs, guillemots can safely occupy even the most precarious sloping ledges.

Joining Ken's classes did far more than expand my theoretical knowledge of ornithology. We also made trips to the Tees Estuary, where I could finally meet the living birds that I had only seen before in books or in Nelson's collection. Here I learned the arcane mysteries of winter wader identification, the subtle clues that distinguish knot from ruff from dunlin – to the uninitiated all just grey specks on distant mudflats. By showing me where and how to look, Ken opened my eyes to an untapped treasure trove of nature, even in this landscape of belching chimneys and smouldering blast furnaces. But I soon discovered a great benefit of living in Middlesbrough for a budding naturalist. It lies surprisingly close to truly wild lands. The North York Moors rise only a few miles to the south of the eructating morass of steel and concrete, and a little further away to the west lies the land of three rivers – the dales of the Tyne, Wear and Tees. It was Ken who introduced me to this very different world of the Yorkshire and Durham Dales.

The Tees rises at the head of one of these dales, all of which are wild and remote places. It's hard to imagine more of a contrast to the tangle of industry that hems in the river at the other end of its journey to the sea. One of our trips was to see black grouse performing their spring dance spectaculars, which meant leaving Teesside at two in the morning – a proper adventure for a young lad. The early start was because we had to arrive at the 'lek', as the grouse's dance ground is called, well before dawn, before the birds begin to assemble. Even before it was fully light, a strange hissing and bubbling began to drift over the grassy moor. In the grey glimmer of pre-dawn, I was able to make out one dark shape after another, parading up and down, bowing and cooing.

As the monochrome world of night began to acquire some colour, I could see about eight or nine birds – males – each jet black but looking remarkably like someone had stuck large white chrysanthemums up their bottoms. These bright-white under-tail feathers really stood out in the pre-dawn light. Each male was defending his own tiny patch of grass on the lek, with the most desirable patches, and so the hardest to defend, located at the centre. That's how a watching female knows which males are the fittest and will therefore make the best fathers for her chicks. But in the half-light of dawn, the females, called grey hens, were almost impossible to see. I did eventually spot one skulking through the grass, and then, as the sun rose over the horizon, all the performers melted away and what was an enchanting scene became just another bit of grassy moorland, indistinguishable from the rest of the vast and rolling landscape that was now illuminated in golden hues.

It was still only 5 a.m. and there were plenty of other birds to see and hear. The shrill piping of a common sandpiper, objecting to my intrusion into its world, and then the bird itself, flying off along a moorland stream on stiffly bowed wings. The cascading songs of meadow pipits as they parachuted from on high in their display flights. Cuckoos calling from the valley below. It was entrancing, and being out and about this early made me feel as if the members of our small party were the only humans on the planet. It was a privileged glimpse into a more vibrant world. By the time most people were rising from their beds, the day already felt old, the light taking on a used and dusty aspect and the profusion of life now hidden and silent.

Ken also took me to my first seabird colony, not at Bempton Cliffs, where Nelson had obtained his guillemot eggs, but on a small group of islands about a hundred miles north of

Middlesbrough – the Farnes. We arrived at the small port of Seahouses, on the Northumbrian coast, to catch a boat over to the islands on a day so foggy we could hardly see the entrance to the small harbour. But it didn't stop the boat heading out – the crew knew this area so well they could have done the journey with their eyes closed. Yet I was overjoyed that the weather was as bad as it was. It made my first visit to a seabird colony all the more magical. As we left the harbour, all sound was muffled and deadened by the fog, and the sea was eerily calm. The noises of the harbour, of civilisation, soon vanished. After about ten minutes, new sounds began to penetrate the grey veil – disembodied grunts and shrill screams, slowly building in intensity. It was as if we had passed between worlds, from the everyday one into an enchanted realm hidden behind curtains of mist.

Then, like a cannonball fired from an unseen ship, a guillemot burst through the fog on whirring wings, saw our boat and swerved, before disappearing back into white nothingness – the first I'd ever seen in the flesh – or perhaps that should be 'in the feather'. While I was still digesting this experience, the shape of the islands emerged from the fog, and I realised that the air – and the sea – was full of guillemots, razorbills and puffins. And it smelled stronger than Teesside – a pungent aroma so powerful I could both smell and taste it. Smell memories seem to be stored deep in the brain, ready to pop up and remind you of past events, no matter how long ago, and I still get the same thrill as on that day whenever and wherever I get a whiff of seabird.

We landed on both Staple Island and Inner Farne, the latter covered in a nesting colony of Arctic terns which mounted a vigorous defence of their eggs and chicks with chattering dive-bombing runs and sharp beaks and claws that drew blood from any unprotected head. Scattered over the island were the nests of eider ducks, the epitome of calm in comparison to the terns. No matter how close you got, they sat tight on their eggs, which nestled snug on a thick layer of eider down. Local legend has an explanation for this extreme tameness. Along this stretch of coast, eider ducks are known as Cuddy ducks, an abbreviated version of Cuthbert. They're named after Saint Cuthbert, a seventh-century monk who spent ten years as a hermit on Inner Farne. He's said to have protected the ducks and tamed them so that they nested right up to the walls of his cell. He obviously did a good job, as they're still very tame and continue to nest up against the walls of the chapel that replaced Cuthbert's cell in the fourteenth century.

On Staple Island, shags were nesting on the clifftops, reptilian in their scaly green feathers. Fulmars swept through steep-sided gullies on stiff wings, and the triumphant calls of kittiwakes mingled with the roar of the sea as it thrust into narrow clefts in the cliffs. I have no idea how long we spent there. It really did feel like a world apart, one where time stood still and where I was fully absorbed in the present moment. Over the years, I've been to other places where nature is so enthralling that it's like being in a trance – among the lesser flamingos in Kenya's Rift Valley or the tens of millions of wintering monarch butterflies in Mexico's Sierra Madre Mountains. But the secret world of dawn in the high dales and the throng of seabirds on the Farnes were truly life-changing experiences. If I hadn't been hooked before, I was now.

And thanks to Ken, I extended my passion beyond the British Isles. He took me on my first trip abroad, to Tangier in Morocco to watch the spring migration over the Strait of Gibraltar. The air was full of birds I would have been unlikely to see in Britain – booted, Bonelli's and short-toed eagles, black kites and white storks (and a few much rarer black storks). If my trips to the dales and the Farnes had given me an appetite to see more of Britain's natural history, my sights were now firmly fixed on the whole planet. I knew I wanted to spend all my time in pursuit of nature, and I felt that what this plan needed first was some academic qualifications. I chose the University of Bristol in part because of its academic reputation but also because Bristol isn't far from Slimbridge on the Severn Estuary, the headquarters of the Wildfowl and Wetlands Trust, set up by Peter Scott (originally as the Wildfowl Trust) in 1946. The reserve attracted winter flocks of Bewick's swans and white-fronted geese, which were not regular visitors to Teesmouth, but there were also flocks of golden plovers, lapwings and other waders which were familiar to me. It was a birdwatching home-from-home – and, after all, I needed something to do when I wasn't studying.

It later proved to a be a wise decision since the University of Bristol was virtually next door to the BBC in Bristol, home of the Natural History Unit. In the mid-1980s, I managed to worm my way into a job there, which eventually took me all over the planet, fulfilling the dream that had shaped itself as I'd watched thousands of birds drifting across spring skies from Africa to southern Spain. But all my family remained in Middlesbrough, so I've never lost my connection to the North-East, and I've continued to explore my old haunts along the Tees, from its mouth to the high dales, places where my joy in the natural world was born and nurtured. This is why, in the winter of 2017, I sit huddled against an icy blast whipping through the hide as a black-necked grebe, impervious to the cold, parades itself for an appreciative audience of birdwatchers.

As I watch it, my mind drifts back over those early experiences that came to shape my whole life and I realise how important this river and the lands that it drains have been to me. It feels like the right time to make a journey along the whole length of the Tees, a kind of homecoming after travelling over so much of the world. My experiences of the natural world and the problems that it faces around the globe have given me a renewed appreciation of the natural world of the Tees Valley. I see it now in a new light and in far broader contexts – geographical and historical, economic and political – than I could ever have imagined in my youth.

The Tees is not the most famous river in Britain – a great many people I meet have never even heard of it and certainly couldn't point to it on a map. Yet it deserves to be better known. It might be only eighty-five miles long, but it crashes over one of England's biggest waterfalls, at High Force, and upstream from this it cascades over what's often described as England's longest waterfall, a 600-foot cataract known as Cauldron Snout.

The Tees rises high on the flanks of Cross Fell, the highest point of the Pennines and only about 250 feet lower than Scafell Pike, England's highest mountain, just thirty-four miles to the south-west as the crow flies. A long-distance walk, the Teesdale Way, follows the river for most of its length, and that will be the route I'll take over the following chapters. But rather than follow the flow of the river, I'll make my journey from mouth to source, to contemplate the natural and social history of this North Country river as I go, and how that reflects on stories from the wider world. In travelling the Tees, I'll also be tracing a journey back into my own history, as several of my ancestors were dalesfolk, living in and around Middleton-in-Teesdale before moving downstream and eventually getting swept up in the economic gravity well of the docks and industry that gave birth to Middlesbrough in the first half of the nineteenth century.

So, forsaking the relative luxury of Saltholme's hide, I retire to the RSPB's excellent café above the visitor centre to plan my venture. Six months later, when the weather is at least a little warmer, I'm back in the North-East to begin my exploration of the Tees at the farthest point from its source on remote Cross Fell, the end of the South Gare breakwater. The Teesdale Way ends (or, in my case, begins) at the lighthouse on the tip of South Gare, a two-and-a-half-mile wall projecting out into the North Sea from the southern bank of the river-mouth. It was built between 1863 and 1888 to protect ships as they entered the river. On the other side of the estuary, the shorter breakwater of North Gare was completed a little later, between 1882 and 1891. Together, they guard the entrance to the river.

I owe the title of this chapter to Roger Jones, a friend and colleague from my days at the BBC's Natural History Unit in Bristol. We worked together on a series called *The Living Isles* (1986), about Britain's natural history from the end of the last Ice Age to the present day. Roger, who is now a painter and poet, came up with a succinct phrase to describe how at the end of its journey, a river gradually mingles with saltwater until finally it becomes the sea. It passes, he said, from senility to salinity.

This mixing of waters, which pulses up and down the estuary twice daily with the tides, creates productive conditions. Myriad creatures hide in the mud and sand, and fish swarm in the waters, together feeding a great variety of birds. Despite the effects of industry over the years, the estuary is still busy with natural life. But if I could see this place as recently as 1800, it would be unrecognisable. I would be hard-pushed to notice many signs of development. Industry once flourished here, in medieval times, when salt was produced near present-day Cowpen Bewley, Greatham and Saltholme and exported along the river. Salt extraction here may even go back into pre-history, but these industries had largely vanished by the sixteenth century.

As the nineteenth century dawned, only a few small ports and fishing villages lay scattered along the riverbanks, although plenty of ships navigated the ever-shifting channels and sandbanks to reach the larger ports of Stockton and Yarm further upriver. Yarm had been an important port since the Middle Ages, but by the nineteenth century Stockton was growing in importance as the Industrial Revolution increased the demand for coal from County Durham's mines.

There were also some large estates and manors scattered along the banks of the lower Tees. In the seventeenth century, William Hustler, a cloth merchant from Bridlington, bought

Acklam Manor since he wanted a place close to the river from which he could trade. A map from this time even shows a dredger operating on the river, creating safe passage for merchant vessels along an already-busy Tees. Middlesbrough itself, however, was little more than a farm, home to no more than twenty-five people. Scattered around were small villages. Their names – Ormesby, Thornaby, Tollesby, Stainsby, Maltby – reveal their origins as Viking settlements: the suffix '-by' means 'village' or 'farm' in old Norse. Yet the first settlement that would eventually become Middlesbrough pre-dates even these Viking settlements.

In AD 686, after leaving his sanctuary on the Farnes, Saint Cuthbert consecrated a monastic cell here for Saint Hilda, abbess of Whitby, a town on the coast about twenty-five miles away. By the twelfth century, this had become the Church of St Hilda of Middleburg: 'Middle' because it was around the middle of the important Christian route connecting the abbey at Whitby with the cathedral at Durham, where Cuthbert's earthly remains now rest. Before being dissolved by King Henry VIII in the middle of the sixteenth century, the Benedictine priory that was established at this significant location was of national importance. However, although it lasted for 400 years, today not a single trace remains.

Apart from this singular claim to fame, history has largely bypassed this obscure corner of the North-East – at least, it did until the first half of the nineteenth century. At this time, just a little way upriver, the Industrial Revolution was, quite literally, building up a head of steam. In 1825, steam locomotives rolled along the Stockton and Darlington Railway, the world's first public railway system. They hauled coal from the Durham collieries around Shildon to Stockton, on the Tees. At the time, Stockton-on-Tees was the lowest bridge across the river and consequently was an important port. Coal was loaded onto ships here and transported around Britain by sea to fuel expanding industries. Georgian industrialists saw the potential for making money and invested heavily in the area – and one of these entrepreneurs would transform the fortunes of the little hamlet of Middlesbrough.

Joseph Pease was a Quaker, coalmine owner and banker, and an important investor in the port of Stockton. He and his fellow investors had already spent a tidy sum in straightening the Tees downstream from Stockton, on the assumption that Stockton would remain a key transport hub. But ships were growing in size to cope with the increased demands of burgeoning industries around the country, and it soon became clear to Pease that, to accommodate these ever-larger vessels, he needed a deeper-water port, closer to the river-mouth.

In 1828, he took a boat downriver to examine locations for a proposed extension to the Stockton and Darlington Railway. His diary entry for 18 August, while he surveyed Middlesbrough Farm, reads, 'imagination here had ample scope in fancying a coming day when the bare fields we then were traversing will be covered with a busy multitude and numerous vessels crowding to these banks denote a busy seaport'. The railway came, and in a short time that imagined bustling port became a reality. Initially christened Port Darlington, a new town sprang up to house the dockworkers. It was laid out in a grid-iron pattern between the south bank of the river and the railway line and by 1851, Middlesbrough's population had risen from 25 to 7,600.

Easy access to fuel from the Durham coalfields meant that Middlesbrough soon became much more than just a port. It became an industrial boomtown, as brickworks, potteries and even a sailcloth company moved in. But industrial development went into overdrive when Joseph Pease offered a site on his new property to the industrialists Henry Bolckow and John Vaughan, on which they built a rolling mill and foundry to work iron. They used the port and the railway to transport iron ore from around Whitby (nearly thirty miles by water along the Yorkshire coast) to furnaces near Witton Park in County Durham. The resulting pig iron (crude iron as it emerges from a furnace) was then transported back to the foundry at Middlesbrough. It didn't take the two shrewd businessmen long to realise that working pig iron imported from elsewhere was not economical, so Bolckow and Vaughan built a blast furnace on the banks of the Tees to produce their own iron from imported iron ore. It was a good idea. By 1868, there were over a hundred furnaces lining the river between Stockton and Redcar, belonging to many different companies. Middlesbrough was now the most famous ironmaking town in the world and was dubbed 'Ironopolis'. In the 1870s, Middlesbrough produced 30 per cent of the UK's pig iron and 15 per cent of all the pig iron in the world.

So, let's hear it for Middlesbrough. It may not be as famous today as in the days of Ironopolis, yet it's immortalised worldwide in its steel girders. I once caused a small tailback on the Sydney Harbour Bridge by driving slowly enough to read the words 'Dorman Long, Middlesbrough' embossed on the girders. The same wording is embossed on trendily exposed girders in Brewhouse, my favourite pub in Bristol (although that's not the only reason it's my favourite pub). Dorman Long was one of the surviving companies from the height of industry in the nineteenth century and absorbed the old firm of Bolckow and Vaughan, along with several others. Steel from Middlesbrough also spans the Limpopo, Zambesi and Nile Rivers in Africa and makes up the Qiantang River Bridge in China, the Storstrøm Bridge in Denmark and, oh yes, the Tyne Bridge. Newcastle's iconic river crossing was born in Middlesbrough. And, just to rub it in, the Angel of the North, Anthony Gormley's famous statue that overlooks the A1 near Gateshead, was also made on Teesside. On the inside, it bears the inscription, 'Built for Geordies by Teessiders'.

When I was young, iron and steelmaking on Teesside had been further amalgamated into the British Steel Corporation, nationalised in 1968 under Harold Wilson's government, and there were many fewer furnaces along the river. The three furnaces at Clay Lane, however, were visible from where I lived on Priestfields Estate. If the furnaces were tapped at night, particularly a cloudy night, the sky glowed orange, like some manmade volcanic eruption. When, as seemed obligatory in teenage years, I read Tolkien's *The Lord of the Rings*, my mental image of the fires of Mount Doom in Mordor was always coloured by the sight of the nightly eruptions from the Clay Lane furnaces. Years later, I spent some time working on those furnaces, as a general labourer, to earn money during the long summer breaks from university, at which time I got a close-up view of Mordor.

In very simple terms, a blast furnace uses a blast of superheated air to react iron ore (iron oxide along with various impurities) with coke, made from coal in nearby coke ovens, and limestone. As the coke burns it produces carbon monoxide, which reacts with the iron ore to release metallic iron and carbon dioxide. Some of the carbon dioxide along with any impurities from the ore react with the limestone to produce 'slag', which in large part consists of calcium silicate. When the furnace is 'tapped', molten iron pours into a huge submarine-shaped structure – a torpedo ladle – parked on railway lines beneath the furnace, and the slag runs through channels on a raised floor that surrounds the furnace, and into bell-shaped ladles parked around its edges, for later disposal.

My job was to clear out the slag channels after the furnace had been tapped and the remaining molten slag had solidified. I also had to clear up solidifying iron and re-lay melted railway tracks on those remarkably frequent occasions when the cascade of molten iron partially missed the hole at the top of the torpedo ladle. When it did, it produced a convincing imitation of a real volcanic eruption, but I knew that after the firework display was finished, my labour gang would have a hot and dirty job to clear up the mess. For reasons I've yet to fathom, I was also detailed on a regular basis to whitewash the walls of the furnace floor. They stayed white for about as long as it took the whitewash to dry.

During this pointless exercise I discovered that nature is ever resourceful. I occasionally noticed movement in the thick metallic dust that covered everything here, including us labourers. So, out of sight of my gang supervisor, I started searching through the dust and eventually discovered a house cricket. These creatures are not native to the British Isles and can't survive year-round in our climate, but they thrive around artificial heat sources. In the past, they were the 'crickets on the hearth', living in stonework around fires. They were common in bakeries, where their trilling songs were a familiar accompaniment to breadmaking. But I was impressed by the resilience of these little crickets, dodging the lava flows of slag, and the heat that melted the soles of my boots if I stood in one place too long – and I guess all they had to eat were the scraps that fell from the lunch boxes of workers.

Years later, I returned to Teesside with a film crew to capture my discovery for *Alien Empire*, a series on insects that I was producing for BBC One. By then, the Clay Lane furnaces had closed and been demolished, but a new furnace, the largest in Europe, had opened at Redcar, and loomed over the South Gare breakwater. It was a lot more hi-tech than the old furnaces

I'd worked on, and nobody was whitewashing the walls, but the crickets had moved in. Today, in an era of climate change, helped along by all that carbon dioxide billowing from blast furnaces, house cricket colonies survive more successfully outside, and, on places like landfill sites, where decomposition provides added warmth, they can now even survive the winter without need for the fires of Mordor.

As a waste product, slag is simply dumped into vast slag lagoons, where it cools and solidifies. It might seem an unlikely comparison, but these places remind me of the slopes of Kīlauea on Hawaii's Big Island. This active volcano spews out lava of a thick, ropy texture called 'pahoehoe', a Hawaiian word which has now been adopted by geologists worldwide to describe this type of lava. Pahoehoe lava solidifies into elegant swirls and textured patterns, and newly cooled slag looks very similar. On Kīlauea, cooled lava flows are soon invaded by lava crickets which move in ahead of any other living organisms, when the lava has barely cooled. They feed on other insects that are accidentally blown onto the barren lava fields and perish. So, in the end, it's not so surprising that I found its kin thriving around a man-made volcano on the other side of the world.

Just as on Kīlauea, the flows of slag are eventually colonised by plants. South Gare is made up of thousands of tons of slag which were dumped to create the breakwater. Some areas, like Cabin Rocks, still resemble a moonscape of broken rocks with little soil, although in spring the bare slag is covered by carpets of yellow kidney vetch and bird's-foot trefoil. The only other place that I've seen such displays is on the machair, the remote coastal grasslands facing the wild Atlantic in the Outer Hebrides, a stark contrast to the view from South Gare. However, the lunar landscape of Cabin Rocks does have its own attractions. It's an excellent place for two scarce butterflies – the dingy skipper and the grayling. Likewise, although you're most likely to find proliferous pinks in the dry Brecks of East Anglia, the nearest thing we have to a desert in Britain, the barren slag of Cabin Rocks suits them just as well. Other areas of slag have developed an even more diverse flora. While slag is rich in calcium, it's poor in other nutrients and doesn't retain moisture, so is a perfect habitat for plants adapted to dry chalk and limestone grasslands. Bird's-foot trefoil, kidney vetch, fairy flax and yellow-wort all thrive on these slaggy grasslands, plants that I'm very familiar with amid the more conventional beauty of the Cotswolds or the South Downs.

Although this peninsula of slag was built to protect shipping entering the estuary from the rage of the North Sea, it's also a very good birdwatching spot. It sticks some way out to sea and so gives a good vantage point to watch birds out over the open water. In summer, common terns fish the brackish waters and ferry their catches back to their chicks, raised on islands in pools like those at Saltholme. Among them are Arctic terns, hard to tell from common terns at a distance, hence birdwatchers often refer to these unknown birds as 'comic' terns. The nearest nesting colony of Arctic terns is on the Farne Islands, where I first met these birds, but a hundred miles is nothing to a bird that each year flies to the Antarctic, to spend our winter in the perpetual daylight of an Austral summer. Sandwich terns are much bigger, and easier to spot, with their heavy black, yellow-tipped bills. Occasionally, roseate terns also visit the estuary. Currently, their only nesting colony in the UK is on Coquet Island, off the Northumberland coast about twenty miles south of the Farnes, so an easy flight to Teesmouth on those long, elegant wings.

Like common terns, the appropriately named little tern – the smallest of our terns – also nests around the estuary, but in much smaller numbers. It's now one of our rarest seabirds and nationally its population is still declining. Little terns like to nest in sand dunes or on open beaches just above high water, and for the last 100 years at least, there was a substantial colony in an area known as the Ducky, at the top of Coatham Sands, the beach that runs along the North Sea coast between South Gare and Redcar. But this site is heavily visited by people and little tern eggs are laid in nothing more than a scrape in the sand, which makes them very vulnerable. So, despite fencing and wardening, the Coatham colony has dwindled over the years. There have been attempts to lure the terns back by using sound recordings and decoys, painted to look like terns by local school children. It has proved a great way of alerting these kids to the national treasures on their doorstep but, sadly, few little terns have returned.

In the mid-1990s, most of the Coatham colony moved to a less-disturbed site at Crimdon Denemouth, about six miles to the north, on the coast of County Durham. In 1998, this was the most prolific colony in Britain, but in 1999, the entire colony was destroyed by a single egg collector. In the following year, not surprisingly, no birds returned to nest here. A few went back to their old colony at Coatham, but continued disturbance there meant that no chicks at all were reared in 2000 either. Not a great start to the new millennium. In 2001, birds began to return to Crimdon, and numbers gradually built back up. In 2016, eighty-two birds nested there and reared at least fifty-eight chicks. However, in 2019, the little terns relocated closer to the Tees, to Seaton Beach, just to the north of the river-mouth. Seaton Beach is much more heavily used by people, so this seemed like a bad idea on the part of the terns. However, they must have known what they were doing since, following the move, they've had several good seasons, although vigilant wardening is needed on this busy beach. Efforts continue to coax at least some of these birds back across the river to South Gare, to safely fenced areas where, in 2015, one pair nested. Further success would turn this stretch of coastline into a national hotspot for the little tern.

Across the estuary from South Gare is North Gare, a shorter breakwater that marks the northern extremity of the Tees Estuary. It's only a short flight across the river for a tern but a longer trek for me to complete the first stage of my journey along the Tees by exploring the furthermost reaches of its estuary.

The lowest crossing point on the river is the Transporter Bridge, six miles upriver from the mouth. This extraordinary structure, which is the longest remaining transporter bridge in the world, has become an icon of the town. In 2002, it was central to the plot of the third series of the comedy *Auf Wiedersehen Pet,*[*] in which a group of Northern working lads

---

[*] The original idea for this successful series – of workers from Thatcher's Britain of the 1980s who are forced to seek employment abroad – came from a bricklayer in Stockton-on-Tees.

take on the job of dismantling the Transporter and rebuilding it on an American Indian reservation in Arizona. If you have to ask, you need to see the series. The visual effects were so convincing that the BBC, in an explanatory caption at the end of the series, felt it necessary to reassure the good people of Middlesbrough that this symbol of their town was not actually being shipped off to America – although it was hardly necessary. This mighty bridge is clearly visible from all over the town.

The Transporter was designed to avoid impeding the ships travelling up and down the busy river and consists of a girder structure spanning the river 150 feet above the surface, supported on either bank by massive pylons. Slung beneath the bridge is a gondola that trundles back and forth on rails, carrying cars and passengers across the river in under a couple of minutes. It's not a hard concept to grasp, but it famously fooled comedian Terry Scott, who mistook it for a conventional bridge (how?) and drove his Jaguar off the edge of the Middlesbrough side on his way to a performance in Billingham on the other side. Luckily a safety net is slung below the end of the road and Scott was unharmed, although he certainly made a lot of Teessiders laugh – if not in the way he was expecting.

The bridge was opened on 17 October 1911 by Prince Arthur, Duke of Connaught, after which a large party of top-hatted and be-wigged gentlemen crowded onto the gondola for its inaugural journey. In 2011, for its centenary, the gondola was replaced with a more modern-looking one, but unfortunately just eight years later the bridge had to be closed. Surveys revealed that one of the great pylons supporting the superstructure was sinking into the mud on the riverbank. The repair costs will probably run into tens of millions, so perhaps the bridge has carried its last passengers. Certainly, some of the leaders of Middlesbrough's council feel that the Transporter may now never re-open. Maybe it's time to talk to some American Indians.

Returning to 2017, when I began my journey along the

whole length of the river for this book, the Transporter is still operating, and it carries me across to Port Clarence in no time at all. The drive from Port Clarence to North Gare takes in the whole gamut of modern industry on the Tees: chemical works, oil terminals, gas processing plants, a nuclear power station and a huge oil rig, in the process of being dismantled. The latter structure is so impressive I pull over to gawk at its sheer size, although I imagine it felt a lot less imposing when it was out in the North Sea, facing a winter storm.

The turn-off to North Gare takes me onto Teesmouth National Nature Reserve and initially across rough grazing pastures leading up to the car park. The fence poles along the road provide song posts for redshanks and, as I pass, one takes off, flashing its bright-white rump and trailing wing edges while piping a loud warning. This bold pattern makes redshanks one of the easier waders to identify, even in winter. Although many of the waders that make Teesmouth famous for its birdlife are winter visitors, redshanks are here year-round, breeding on these rough pastures. In spring and summer, their fluting songs are one of the most evocative sounds of the estuary, and in winter, their numbers are vastly increased by visitors escaping more northerly climes.

It's a short walk from the car park to a track running around the damp grassland behind the dunes that fringe the north shore of the river-mouth. In midsummer, Teesmouth is a much more colourful place than it is in the grey depths of winter. Hundreds of northern marsh orchids grow in this wet grassland, lush spikes of a deep and intense mauve. Northern marsh orchids are often big and succulent-looking, but I soon spot some particularly tall and exuberant specimens. They're also paler in colour and with a slightly different shape to the lower lip of their flowers. I suspect these are hybrids with the common spotted orchid, their large size down to hybrid vigour, a phenomenon, more properly called heterosis, that causes hybrids to be more robust or tougher than either parent.

These orchids, belonging to the genus
Dactylorhiza, are notorious both
for hybridising and for confusing botanists. The idea of just
what constitutes a species and
exactly which species occur in
Britain has changed many times
over the years, which means that
if you were botanising in western Scotland a few years ago, for
example, you might have found yourself
being thrilled to find a western marsh orchid, *D. majalis*. But
today, you'd have to console yourself with finding nothing
more than a subspecies of northern marsh orchid, the same species as those growing round my feet at North Gare. *D. majalis*
itself is no longer regarded as a British species. For any orchid-
twitchers, though, there is a unique variety of northern marsh
orchid growing here. *Dactylorhiza purpurella* var. *atrata* grows
only on the dune grasslands running up to North Gare, and a
particularly beautiful orchid it is, much more intense in colour
than normal northern marsh orchids, although the standard
northern marsh orchid is certainly colourful enough.

The only real way to sort out the confusing mess of
Dactylorhiza orchids is to look directly at their DNA, which
has recently been done. It was a revealing exercise which
showed that many of our different marsh orchids had their origins as hybrids in the distant past, including the northern marsh
orchid. This species arose millennia ago as a hybrid between
the early marsh orchid and the common spotted orchid and is
now its own *bona fide* species. So, in finding this recent hybrid
at North Gare, I'm really witnessing evolution in action. Who
knows when another one of these hybrids will establish itself
as a species in its own right?

For hybridisation to happen, something must carry pollen
from one species to another, and I soon spot a possible culprit

– a narrow-bordered five-spot burnet moth, a mouthful of a name for a spectacular creature. It's a day-flying moth, happy to buzz around flower meadows in bright sunshine, showing off iridescent green wings marked with red spots – five of them, of course. But it's very hard to distinguish this moth from the five-spot burnet, a different species with a confusingly similar name. I'm assuming that this is a narrow-bordered five-spot since the other species is much rarer, and confined to southern England. However, because they're so hard to tell apart, it's almost certain that we don't have an accurate picture of the distribution of either species, so I should probably pay this individual a little more attention.

Closer scrutiny is, however, out of the question. I'm soon distracted by swarms of these colourful moths, just begging to be photographed. They love to sup nectar from the flowers of thistles and knapweeds, and the marsh thistles growing a little further along the path are covered in burnet moths, more than a dozen on each head jostling for the best position. Time to settle down with my camera.

Along the seaward edge of these glorious wet grasslands are lines of dunes that mirror those on the southern side of the estuary. After North and South Gare were built, extensive dunes grew up as sand was blown against the walls of the breakwaters. Windblown sand is soon anchored by the extensive roots of lyme grass and marram grass, which stabilise the dunes and allow other species of plants to gain a foothold. One of the most spectacular here is purple milk-vetch, whose densely packed flower heads straggle through the marram grass. It also grows on the opposite shore, on the slag of South Gare. Generally, it's a plant of chalk and limestone, but in southern England its populations in the Cotswolds and Chilterns and on the dry Brecks of Norfolk have declined, largely owing to the spread of more intensive farming. In Britain as a whole, it's now considered an endangered plant, but it occurs in some abundance along the north-eastern coast of England and the eastern coast of Scotland, and here at Teesmouth it forms spectacularly large patches that really catch the eye when lit by the sun.

In that same brief patch of sunlight on an otherwise-stormy summer's day, I notice a stretch of the dunes glowing a vibrant yellow-green. A closer inspection reveals a vigorous clump of twiggy spurge. It's not a native plant and in some parts of the world it's become a real pest, invading pastures and turning them bilious yellow, although here, in among dune grasses, it seems better behaved, and confined to this one luminous clump.

From the top of the dunes, I can look across to South Gare and see the sweep of the southern shore of the Tees, lined with industry and urban development. I can also see back across the grazing marshes and mudflats that are still such ecologically vibrant places, despite so much heavy industry. I'm grateful that I first began to learn the joys of natural history in this very unnatural landscape. Most of us instinctively draw a clear and sharp distinction between the natural world and the human one, between the countryside and the city, between nature and un-nature. But the modern world isn't like that. The hand of humanity is everywhere, as we'll see as our journey unfolds. Crops and grassy pastures are just as obviously human arte-facts as steelworks and chemical works, but what about those wild moorlands around the head of the Tees where my journey will eventually take me?

They were once forested, before our ancestors began clearing the trees for agriculture. And even our remaining forests have been shaped by millennia of use. We don't even know what the 'wildwood', the primeval forests of Britain and Europe, would have looked like, although some ecologists suggest they would have been a mosaic of grassland and open forest, created and maintained by large herds of big grazing animals, nothing like the impenetrable tangle of fairy tale and legend and nothing like woodland nature reserves that are today scattered across the country.

Deep history teaches us the value of questioning our

assumptions about nature. What is natural? How do we define wild nature? In the United States, wilderness is actually defined by law in the Wilderness Act of 1964: 'A wilderness, in contrast with those areas where man and his own works dominate the landscape, is hereby recognized as an area where the earth and its community of life are untrammelled by man, where man himself is a visitor who does not remain.'

The Act currently protects over 100 million acres of land, most of which is no more wilderness than the industrial scene I see before me. America's original inhabitants, some 500 nations of American Indians, lived off this land for many thousands of years and left their mark. Admittedly, they had less of an impact on species numbers and diversity than the iron and chemical industries have had on the lower Tees, but their influence was still substantial, although conveniently forgotten or misunderstood by the descendants of Europeans who claimed these territories as their own on the grounds that the American Indians had failed to follow the biblical injunction to plough the earth and subdue the land.

The value of exploring the natural world in the industrial landscapes of Teesmouth is that it leaves you in no doubt that nature has had to fit into the spaces left by the human world. Yet the same is true almost everywhere else in Britain, although it may be less obvious in more rural settings. Across the globe, the story is the same. A stark realisation has emerged from my travels around the world in the decades after leaving Middlesbrough: business as usual is no longer an option. A biodiversity crisis intertwined with a climate crisis leaves me in no doubt that we need to fundamentally change our view of nature and of our place in it. The journey ahead of us is along one river in northern England, but it could be along any river, or through any landscape. If we make that journey with eyes and mind wide open – and we will – the urgency of this paradigm shift becomes blindingly clear.

Any newly emerging worldview must also recognise that in

the foreseeable future we will continue to live in a human-dominated world, which makes it all the more important that we learn to make space for nature with a mindset that accords it far more respect than we give it at present. Friends of mine from American Indian nations have told me that they see nature in a very different way from how it is defined in the Wilderness Act. Rather than dismissing an alternative understanding of the natural world, as has so often happened in the past, our new mindset for the future would do well to incorporate some of these viewpoints. We should also look back to our own past. One encouraging aspect of doing so is that we can see how in the past we've been able to coexist with a much richer natural world, which means that we already know how to achieve this again in the future.

To fully understand the natural history of Britain, we must also understand its human history. As we travel upriver, we'll explore further and further back in time to see how our social history has created our natural history. Again, it's just one river in England's North-East, but the stories that we'll uncover are of much wider importance. The Tees Valley has played an important role in the creation of the country we now call England, its politics and economics, as well as its literature and art. As we walk the Teesdale Way together, we'll become aware that the microcosm illuminates the macrocosm. What we discover beneath our feet and over our heads are not just quirky stories of local interest – although there are plenty of those. These stories reflect much broader concepts that are of national and international significance.

Our journey takes us through just eighty-five miles of geography but through hundreds of millions of years of geology and evolution. Further upriver, our journey will take us back to these most distant times, long before humans were even a twinkle in the eye of evolution, when geological forces on unimaginably immense scales forged the landscapes that we'll pass through. However, most of our historical journey will take place in the 12,000 years or so since the ice sheets of the last Ice Age glaciation released Britain from their frozen grip. This period, known as the Holocene, encompasses our transition from hunter-gatherers to industrialists and urbanites. Those geological forces from much earlier times created the raw materials such as coal and iron ore that helped fuel this transition and create our modern economy. In turn, this has pushed our world into a new geological age, in which a single species now exerts a planetary force as great as any geological process. By extracting energy from fossil fuels and releasing carbon dioxide as well as severely depleting the Earth's biodiversity, we've profoundly altered the planet's ecology. We've pushed the climate out of its stable Holocene state and into an entirely new and less stable one. We have entered a period that some scientists now call the Anthropocene.

The dating of the geological periods that we'll encounter in our travels in the following pages is necessarily vague – they're so remote in time that their boundaries can only be defined as somewhere within a few tens of thousands of years. But we can date the start of the Anthropocene precisely: 1950. This was a time when social and economic conditions aligned to create what economists and climate scientists now call the 'Great Acceleration' – when industrial development, consumerism and capitalism all went into overdrive. As we're about to discover, this period had massive effects on the Tees. The Great Acceleration, though, had its roots in the industrial and agricultural revolutions of the eighteenth and nineteenth centuries and, along the Tees, we'll also find ample evidence of the effects of these processes on both nature and society. In fact, some scientists and economists prefer to date the start of the Anthropocene to the Industrial Revolution. Yet others see the start of this period as lying much earlier in time, since, even early in our evolution, we humans soon began to influence the environment in more dramatic ways than most other species. Over the course of our journey, we'll have cause to consider all these periods of history and how they have contributed to the world in which we now live.

There's no better place than the Tees Estuary to explore what the Industrial Revolution in the nineteenth century and the boom times of the early twentieth century meant for both nature and people across the whole country. The next stage of our journey will take us from a river teeming in fish and supporting thriving fisheries to one that was all but dead – and along the way, we'll discover some additional truths in Karl Marx's observation that 'all great world-historic facts and personages appear, so to speak, twice... the first time as tragedy, the second time as farce'.* Or, to put it another way, history repeats itself – it has to, since nobody listens.

---

* Karl Marx. *18th Brumaire of Louis Bonaparte*. Joseph Weydemeyer. 1852.

2

Between the Tides

*Seal Sands to the Tees Barrage –
Salmon, Seals and Mud*

Big Salmon once swam in the clear crystal Tees
Now no pink dorsal fins show
Though the powers of man brought the rain to its knees
On Tees banks the flowers still grow

VIN GARBUTT,
'The Valley of Tees', 1972

It's a short drive from North Gare to a small car park just south of Greatham Creek, a broad, muddy channel that flows into the Tees across Seal Sands. Slowing as I cross the road bridge over the creek, I can see the bulky shapes of harbour seals hauled up on its banks, ten or fifteen of them prone on the mud, heads and tails lifted, looking like sausages curling up on the barbecue. Close to the car park, there's a wooden wall along the bank of the creek, furnished with observation slits that allow good views of these basking seals without disturbing them. And they attract plenty of admirers, since these seals have achieved something of a celebrity status, both on Teesside and nationally. Apart from giving the intertidal mud- and sandflats to the north of the river channel their name, seals are a symbol of the rebirth of life here. They're the first documented example of seals re-colonising an industrial estuary after having been wiped out – along with almost everything else – by pollution and disturbance.

When I left Middlesbrough in the early 1970s, the Tees's main claim to fame was that it was the most polluted river in Britain, and it was frequently described as dead. But reports of its death were greatly exaggerated – to borrow a line from Mark Twain, an author more usually associated with an altogether grander river. Wind back the clock to 1850, several years before a young Mark Twain, or Samuel Clemens as he was then, began his training as a river pilot on the Mississippi. If I had stood on the banks of Greatham Creek back then, I would have seen the south bank of the Tees across nearly four miles of open sand- and mudflats, inundated twice daily by the tides. Upriver, Middlesbrough Docks, not much more than a decade old, would be busy with ships loading coal from the Durham coalfields, and Messrs Bolckow and Vaughan would have built their first foundries, rolling mills and blast furnaces to create and shape iron. But downstream from the new town of Middlesbrough, the Tees still ran free, and it had three or four alternate routes to the sea. Throughout history, the river has favoured different routes as silt builds up and blocks the active channel. In the past, it emptied into the North Sea close to Tod Point, a place now far inland and buried under steel works and chemical factories. It lies over two miles from the present mouth between North and South Gare.

The estuaries of free rivers are dynamic places. The main river channel frequently shifts course so that, seen from above and witnessed over eons, a river thrashes and twists like a demented snake as it switches from channel to channel. But this unpredictability doesn't suit human plans. Since I've already made the unlikely connection between the Tees and the Mississippi, let me invite further ridicule and continue the comparison. Over the years, I've spent a lot of time on the Mississippi, from source to delta, and there's no better place to appreciate the conflict between the will of a wild river and the human concept of a useful river.

Like the Tees, Ol' Man River has changed course frequently over the centuries, but on a much larger scale. One of its old mouths, at the end of what today is the Atchafalaya River, is over 150 miles from its current mouth. The Atchafalaya is presently a glorious backwater, where swamp cypress trees are draped with Spanish moss and rare Louisiana black bears splash through the shallows. But in the first half of the twentieth century, the Atchafalaya River began to capture more and more of the Mississippi's water. The Mississippi was preparing to change course and flow once again down its old channel to the Gulf of Mexico. By then, however, Americans had gone to a lot of trouble to build New Orleans and Baton Rouge along its present course, so there was no way they would allow the current Mississippi channel to become a backwater. In the 1950s, the Army Corps of Engineers built a massive water management system at the junction of the Atchafalaya and Mississippi to control the amount of water flowing into the Atchafalaya and prevent a course change. Although this piece of ambitious engineering almost failed during the catastrophic floods of 1973, it's still working – so far.

The case of the people versus Ol' Man River could stand in for any number of rivers around the world that have been tamed and bent to the will of humanity, a list that certainly includes the Tees. As I stand on the banks of Greatham Creek, transported in imagination to 1850, I'm witness to the dawning of a new era for the Tees. On the far southern horizon, Eston Hills rise beyond the town of Middlesbrough, which is already home to nearly 8,000 people. Eston Hills are an outlier of the North York Moors, which rise to greater heights behind them, and at the start of 1850, these hills were still as wild as the higher moors to the south. In the summer of that year, however, John Vaughan discovered iron ore in Eston Hills.

Up to that point, Bolckow and Vaughan had been importing iron ore from mines elsewhere, but Teesside's fledgling industries were already experiencing economic difficulties. Vaughan's timely discovery changed all that. Within a short time, the first iron ore emerged from the new mines, which would soon burrow for three miles under Eston Hills. As people flocked to work in the mines, tented camps sprang up to provide accommodation. Middlesbrough's 'iron rush' was a repeat in microcosm of the California gold rush, which had begun just the year before in the Sierra Nevada Mountains. And to mark this coincidence, the new settlement that was built to provide Eston's miners with more permanent housing was christened 'California'. California is still part of Eston today, so I'm not the first person to make grandiose comparisons between the Tees Valley and the USA.

The Eston mines fuelled a resurgence of industry. In the second half of the nineteenth century, blast furnaces, foundries and rolling mills stretched out along the Tees. Drawn by the dramatic growth of iron- and steelmaking, shipyards opened and chemical works filled in any gaps with their baffling maze of pipes. Following the iron rush, the unprecedented scale of the expansion of industry here prompted Sir H. G. Reid, author of *Middlesbrough and Its Institutions* in 1868, to write:

> The iron of Eston has diffused itself all over the world. It furnishes the railways of the world; it runs by Neapolitan and papal dungeons; it startles the bandit in his haunt in Cicilia; it crosses over the plains of Africa; it stretches over the plains of India. it has crept out of the Cleveland Hills where it has slept since Roman days, and now like a strong and invincible serpent, coils itself around the world.

Iron and steel built the infrastructure of the expansionary Victorian age and Middlesbrough's role in this was so important that in 1862, William Gladstone, then in his second stint as Chancellor of the Exchequer,* paid the town a visit and proclaimed, 'This remarkable place, the youngest child of England's enterprise, is an infant, but if an infant, an infant Hercules.'

By the 1850s, the Tees had already been straightened between Middlesbrough and Stockton by the digging of a couple of channels that cut off two long meanders. In 1810, at Mandale, a cut just 600 feet long saved a journey of two and a half miles along the river. Later, in 1828, a second channel created a short cut around the meander at Portrack. Before this, it was said that the journey between Stockton and Middlesbrough took longer than the journey between Middlesbrough and London. Even so, beyond Middlesbrough, the shifting channels and sandbanks of the estuary made navigation treacherous. In addition, a six-mile-long sandbank called the Bar lay just offshore, and made entry into the Tees almost impossible at low tide. So, during the 1850s, the Tees Conservancy Board was created by an Act of Parliament and given the job of making the river-mouth safer

---

* Just six years later, in 1868, he would start the first of four terms as prime minister.

for the ever-growing number of bigger and bigger vessels. The creation of North and South Gare increased the depth over the Bar at low tide to a comfortable twenty feet and the main channel was edged with slag to prevent the river doing its own thing. Since this also increased the flow of water, which scoured and deepened the channel, it created a permanent deep-water route for shipping. Beginning in 1853, this main channel was also regularly dredged to maintain its depth. Now the good times could really roll.

In 1858, Joseph Blossom captured the early years of Middlesbrough in a somewhat idealised watercolour, which showed the river busy with ships, under sail and powered by steam, and the distant southern bank lined by the imposing skyline of the new town, dominated by St Hilda's Church, consecrated in 1840 and standing close to site of that ancient cell dedicated by Saint Cuthbert to Saint Hilda a millennium and a half earlier. As industry grew, so, too, did the town – and the population. By the end of the nineteenth century, those twenty-five souls who had inhabited Middlesbrough Farm seventy-five years before had grown to ninety-thousand. Many Irish immigrants followed the work here, although most of the population, at least initially, was drawn from closer at hand. One part of my family left the farming lands around York to seek more steady work and wages as Middlesbrough grew. Some of my other ancestors followed the Tees down from Middleton-in-Teesdale, in County Durham, where they'd been lead miners, exchanging one ore for another. Those two branches, both on my mother's side, met in Middlesbrough with my great-grandparents. On my father's side, the Nicholls were long-time inhabitants of Great Ayton on the banks of the River Leven, a tributary of the Tees that we'll explore later. Nestled tight against the northern slopes of the North York Moors, Great Ayton, known locally as Canny Yatton, is indeed a canny place. Despite being only a few miles from Middlesbrough, it's still a picturesque rural village, but its proximity to the rapidly expanding industries and opportunities along the Tees meant that many of its inhabitants, including several Nicholls, moved to the town in the early days of its growth.

But all those 90,000 people, my own family included, spelled disaster for the river that gave birth to the town. Right up until the 1980s, untreated sewage was pumped into the river – a nutritious broth for bacteria. As they break down organic matter, they use up oxygen and with so much organic matter available, the river was soon devoid of dissolved oxygen, suffocating anything that once lived there. Nor is such pollution a thing of the past. History does indeed repeat itself. In 2022, water companies across Britain were forced to admit to the Environment Agency that they'd released raw sewage into Britain's rivers and seas over 370,000 times during this one year, for a total of 2.6 million hours. In other words, there were over 800 spills somewhere in Britain every day.

Water companies are allowed to release untreated sewage but only during heavy storms, to avoid sewage backing up into homes. Many of these releases, however, were during times when the system should have had the capacity to cope. Despite making huge profits, these separate privatised water companies have all claimed that the sewage processing infrastructure needs major modernisation to cope with demand, which we customers will have to pay for. This has prompted some scientists to accuse the water companies of using illegal practices to create profits and bonuses for executives and shareholders – which, incidentally, come close to the figures that the water companies say we'll have to pay to upgrade our sewage systems. However, the use of such damaging practices is nothing new along this stretch of the Tees.

During the nineteenth century and well into the twentieth, industry poured all kinds of toxic substances into the river, including cyanide. One analysis in the 1930s found that around 1,800 pounds of cyanide were discharged into the river *every day*, along with 4,400 pounds of tar acids – so anything that didn't suffocate was soon poisoned. In the 1920s, large numbers of dead fish commonly washed up on the banks of the Tees, many displaying the bright-red gills indicative of cyanide poisoning. Other chemicals also contributed to the toxic cocktail. Pyridine, a hydrocarbon, is produced in abundance as a byproduct in coke ovens but is also used as an anti-fouling agent on submerged structures, where it works extremely well – by being lethal to all forms of aquatic life, even at low concentrations.

Exploring the estuary in the early 1970s, I came across streams that might once have bubbled clear water into the Tees but on occasions now ran with the most extraordinary range of colours – bright orange, bilious green and occasionally vivid blue. I've no idea what was in these streams, but they certainly smelled bad, suggesting it wasn't a good idea to get close enough to find out. In those days, it was almost impossible to imagine that the estuary was once a nationally important fishery for smelt, salmon, sea trout and eels, among others, as well as supporting commercial wildfowlers.

Records of fishing on the Tees go back to the fifteenth century. In the sixteenth century, smelt were so abundant in the estuary that they supported a thriving fishery which soon grew so large that regulation was needed. In 1530, the smelt shoals gained a reprieve when a closed season was declared; smelt could only be caught between 25 April and 1 August, and then only from areas upstream of Saltholme. Yet other regulations were needed to prevent any further escalation of the frequent arguments between those who fished using seine nets* and those using haling nets (framed nets on poles), while other kinds of nets were prohibited entirely. In coastal waters off the river-mouth, crabs and lobsters were also abundant and have supported lucrative local fisheries through the centuries, right down to the present day. When I was a boy, no day out at Redcar was complete without buying fresh crabs straight off the boats as they hauled up on the beach.

By the eighteenth century, the principal fishery on the Tees was for Atlantic salmon. Large numbers were caught at Stockton-on-Tees, enough to supply local needs as well as to sell in the markets of York and Leeds. The Tees became the third-most-important salmon fishery in England, the fish being harvested by both commercial fishermen using seine nets in coastal waters at the mouth of the river and rod fishermen further upstream. In 1867, 10,000 salmon were netted on the Tees. The Fishery Board also issued nearly 1,500 rod licences for sea trout. It was clearly a thriving river.

These salmon and sea trout bred in the faster-flowing waters of the higher reaches of the Tees and its tributaries before migrating out to sea to feed and mature. Sea trout belong to the same species as brown trout, but when young they make the decision to swim downstream and head out to sea, perhaps

---

* A seine net hangs like a curtain in the water, buoyed by floats at the surface and held down by weights along the bottom. Ropes attached to either end of the net allow the sides to be drawn together to enclose a fish shoal.

prompted by a lack of food in the river, perhaps by their genetic inheritance, or more likely by a combination of both. Most brown trout, on the other hand, spend their whole lives in the higher stretches of the river, although some do migrate within the river or from the river to connected lakes. Some, rather unkindly referred to as 'slob trout', even hang about in the estuary, in productive brackish waters.

Young salmon grow in oxygen-rich waters upstream for a year or two, at which stage they're called 'parr', easily recognised by their 'parr marks' – a series of dark smudges down their sides that look like dirty fingerprints. Eventually, they lose their parr marks, turn silver and become 'smolts'. At this stage, they begin their journey downstream and prepare to run to the sea. They find much better feeding at sea than in the river's headwaters, and this allows them to grow to their large adult size and to mature their sex organs, ready for a return, several years later, to breed[*] – usually in the river of their birth. Once upstream again, the females dig out nests, called 'redds', in gravelly areas in which they lay eggs, and males that have grown large and fat on their seafood diet compete to fertilise them.

However, not all parr turn into smolts. The journey downriver is dangerous, the sea is full of big predators that relish salmon, and the journey back upstream is often fatally exhausting. So, some male parr stay in the river. They can't grow to anything like the size of the sea-fed fish, and they certainly can't compete directly with full-sized males, but sperm are cheap to produce in large quantities, so these parr can still become sexually mature. They take advantage of their small size to sneak in under the radar and fertilise a few eggs while the big males are preoccupied trying to outdo each other.

Eels do things the other way around. They feed and grow in

[*] Salmon spend anything from one to four years feeding at sea. Those that return after just one year are called 'grilse'.

the river for up to twenty years, then migrate downstream to spawn at sea. As they mature, their eyes grow larger, and they turn silver along their bellies as guanine crystals are laid down in their skin. When they reach this stage, they're called 'silver eels', distinct from 'yellow eels', their former river-dwelling selves. In the river, an eel can hide in just about any crevice or squeeze into dense vegetation, but in the sea there's no hiding place. The silvering of their bodies is open-water camouflage, used by many kinds of pelagic fish.* The mirror-like crystals reflect the light that penetrates the ocean from above, thus obscuring the silhouette of the fish against the surface from any predators lurking below.

I know all this because back in the 1980s, I shared a lab at Bristol University with an eel biologist. Silver eels are hard to find in the wild because they disperse into the vastness of the ocean, so, at the time, the details of their transformation were still a bit of a mystery. My colleague used various hormone treatments and manipulated environmental conditions to stimulate the eels' Jekyll-and-Hyde act in an aquarium. That meant that I got a privileged glimpse of mature silver eels, very different-looking creatures from the more familiar river-dwelling yellow eels.

For a long time, no one knew where these silver eels went to spawn – they simply vanished into the immensity of the Atlantic. Then researchers discovered that all the eels from Europe, along with those from the East Coast of North America, ended up in the Sargasso Sea – a very strange place indeed. The Sargasso Sea is often eerily calm and in places choked with rafts of a floating seaweed called *Sargassum*. It lies in the western half of the Atlantic, bounded by ocean currents that form a great gyre. It's like the centre of a giant whirlpool, although in reverse. Water piles up in the centre of the

---

* Pelagic fish are those that live in open water, well away from land and from the sea bed.

Atlantic gyre, so, even though it goes against common sense, the surface of the Sargasso Sea is higher than the surrounding ocean. The ocean currents bring the *Sargassum* weed here from the Gulf of Mexico, and it builds up into such huge patches it looks as if the sea has coagulated. I once spent many happy weeks searching these weed beds to film the strange creatures that live only in these floating mats – weird toadfish whose fins look like little clasping hands and sea slugs covered in strange excrescences that make them look exactly like the *Sargassum* weed itself. Beneath the mats, large shoals of young fish shelter from aerial attack, but when the eels arrive here, they shun the safety of the great rafts of *Sargassum*.

Instead, they dive into the dark waters deep below the floating seaweed. Since they spawn at such great depths, the details of their life cycle have been hard to work out, although we now know that the adult eels die after spawning, while their eggs hatch into very un-eel-like larvae called 'leptocephali' (sing. leptocephalus). These delicate-looking, transparent creatures then face a long journey back across the ocean to the rivers of Europe and North America, although they can at least hitch a ride on the currents that create the Sargasso Sea. Even so, those little leptocephali from European eels may take three years to get back; the North American eels, on the other hand, have a much shorter journey. As the leptocephali finally approach the shore after a journey of, in some cases, more than 3,000 miles, they transform into tiny transparent eels called 'glass eels' that, in unpolluted rivers, swarm upstream in unbelievable numbers. Eventually they transform into 'elvers', which look much more like miniature eels, to begin their long lives in freshwater.

As the twentieth century dawned, fish found it increasingly difficult to cross the Tees Estuary in either direction. Those sneaky parr in the clear waters of Teesdale waited in vain for the big fish to return. The Tees Estuary fishery began to collapse in the first decades of the twentieth century and by the end of the 1930s, salmon, sea trout and eels had disappeared completely. And that's how it remained for more than four decades, though no doubt many salmon tried to enter the Tees in this period, since they're instinctively drawn back to the river of their birth after years of feeding at sea.

Salmon are skilled navigators, although quite how they find the right river is still a bit of a mystery. For long-distance navigation they use the sun as a compass, along with an internal clock to compensate for the sun's changing position during the day. They can also see the patterns of polarised light across the sky, which gives them further information about the direction in which they need to travel, and they can detect the Earth's magnetic field. But once close to the river, smell takes over. On hatching, and at several other times in their freshwater lives, young salmon imprint on the particular chemical cocktail of their river, memorising it to guide their return journey several years later.

The sense of smell in both salmon and eels is legendary. Sockeye salmon from the West Coast of North America (Pacific salmon are not closely related to our Atlantic salmon) can detect tasty shrimp extract in quantities equivalent to five teaspoons in an Olympic-sized swimming pool. Another study showed that Pacific salmon can detect extract of sea lion (a major predator of West Coast salmon) at only two-thirds of a drop in that same proverbial swimming pool. I'm still puzzled by how one makes extract of sea lion, but I can't shake the image of a very large blender and a very annoyed sea lion. Nor can I work out what two-thirds of a drop is, since it's still a drop, just a smaller one.* But that aside, it's clear that salmon have a finely tuned sense of smell – and eels may be even better. American eels (which are pretty much the same as our European eels) can

---

* Some sources claim that a drop is a metric unit equivalent to 0.05 milliliters, but actual drop size varies with the surface tension of the liquid in the dropper, so it's somewhat imprecise. I rest my case.

detect one ten-millionth of a drop (or a very, very small drop) of their home water added to that Olympic swimming pool. I can hardly imagine what salmon and eels returning to the Tees must have smelled as industry spread along the estuary. The delicate scents of Teesdale would have been lost in the toxic mix flowing out into the North Sea. Olfactory overload, indeed.

Attempts to rescue the failing salmon industry began in earnest in the 1920s, as calls to control ever-growing pollution in the Tees became louder. The Salmon and Freshwater Fisheries Act was passed in 1923 to protect fisheries across the country but was so full of loopholes in favour of big industry that it had little effect on pollution or on the attitudes of those causing the pollution. One of the first big chemical firms on Teesside, Synthetic Ammonia and Nitrate (later part of ICI), saw the lower Tees as nothing more than 'an industrial stream'. It reasoned that those industries of 'enormous potential development' along the Tees would just set up elsewhere if 'the River Pollution authorities impose too drastic regulations'.* What it wanted was a licence to pollute as much as was necessary in the pursuit of profits.

Even so, a detailed biological and chemical survey of the Tees was suggested, to catalogue the surviving fish and identify the main sources of pollution. The industrialists were not best pleased with this idea. The chief chemist of Dorman Long threatened to resign if the company was forced into contributing to the costs of the scheme. The chemical industries, on the other hand, saw the whole thing as a waste of time, since everyone knew it wasn't the chemical works, busy manufacturing ammonia- and nitrate-based fertilisers, but the iron and steel plants, like Dorman Long, that created the worst pollution. Certainly, waste such as pyridine from the coke ovens that were scattered along the lower river was particularly lethal to

* Quoted in John Sheail. 'A Barrage of Poisonous Water: Inter-War Research on the River Tees'. *Water Policy* 2(4), 299–312. 2000.

fish and marine invertebrates like crabs and lobsters, but the manufacture of nitrates and ammonia can also do plenty of damage. In the end, though, both industries blamed raw sewage and runoff from roads newly coated in tarmac to deflect attention from the damage done by their own effluents – an early example of gaslighting.

Finally, a consortium agreed to fund the survey, although only if it could use its own local analysts. However, in 1926, before the survey had been finalised, there was a dramatic demonstration of the impacts of industrial pollution that was plain for all to see without complex chemical analysis. On 1 May, a General Strike was declared in support of miners locked out of mines as they tried to prevent wage reductions and worsening conditions. Many workers in heavy industry and transport came out in sympathy and soon all industry on Teesside ground to a halt. In September of that year, a 'marvellous recovery in the biological condition of the river' was reported. Many salmon smolts that year made it to the sea and healthy crabs abounded at low tide. However, in December, just days after the coke ovens restarted, dead fish were found on every tide. At the same time, the River Tyne to the north and the Esk to the south, at this time both less polluted than the Tees, were described as crowded with fish.

Despite such clear demonstrations of cause and effect, nothing was done. No one was willing to pay the higher costs of controlling pollution or – to judge from the necessity of the General Strike – of giving their workers a fair deal. The industrialists openly advised the government to turn a blind eye to the effects of unbridled industrial expansion on the local environment and public health lest it should threaten 'what are at present prosperous and populous communities'. Populous, yes – but only the lucky few were prosperous, and they wanted to keep it that way.

By the 1930s, enough scientific information was available to explore ways to reduce the effects of industry on the environment and restore the salmon fisheries, which by that time had collapsed completely. The unashamed response from the industrialists was that 'a scientific solution is not necessarily an economically practical one'. We still hear this today as governments disregard the best scientific advice in favour of their own politically motivated ambitions. Likewise, Teesside's early-twentieth-century captains of industry saw the choice 'between destroying your manufacturers and maintaining the purity of your rivers' as abundantly obvious. Steel and fertilisers should always win out over fish. And so, the Tees continued in its death throes.

It took until the 1970s, when I was getting to know the estuary, before concerted efforts finally began to clean up the Tees, with the eventual aim of restoring the fish runs, particularly those of salmon and sea trout. There was an air of optimism about these projects back then, captured by Middlesbrough singer and songwriter – and acerbic wit – Vin Garbutt. Known as the Teesside Troubadour, he was the embodiment of the cynical humour rife in this corner of the North-East. He was as famous for the hilarious rambling introductions to his songs as for the often-poignant songs themselves. I saw him many times in the early 1970s at folk clubs around Teesside and was deeply saddened to hear of his death in the summer of 2017. He wrote 'The Valley of Tees' in 1972, looking forward to the return of fish, and of nature itself, to the Tees, and at a recent concert introduced the song as follows:

I was optimistic when I wrote this about salmon getting back in the Tees. It was dead – not a living thing in it and here we are now nearly forty years on and there's hardly a Saturday night without someone getting chucked in.

But the optimism in 'The Valley of Tees' was well founded. In 1982, a few salmon were seen, suggesting that conditions in the estuary must have improved, at least enough for some tentative visits. So, in 1991, a million fry and parr from hatcheries on the Tyne were released into the upper Tees, to give nature a head start. Hopefully, these fish would find their way back to a cleaner Tees when they matured.

The partial recovery of fish stocks in the river was the result of a consolidated effort by many organisations to clean up the river. But in large part it was undoubtedly down to the collapse of industries along the river at the same time – in particular iron- and steelmaking. As it did during the General Strike, the shutdown of industry saw an improvement in river conditions, but this time the shutdown was permanent. In 2015, the giant blast furnace at Redcar closed, with a loss of over 2,000 jobs. This was the final blow in a decades-long decline, and it marked the end of large-scale iron- and steelworking on Teesside. Gladstone's infant Hercules grew to maturity and died in just 170 years. The local paper, the *Evening Gazette*, ran the simple headline '1845–2015: 170 Years of Steelmaking Laid to Rest at Teesside'. Vin Garbutt caught the mood of this latest surge in unemployment in another of his songs – 'Slaggy Island Farewell':

> In just one century, you have lived and you have died
> Your birth meant life to many men
> You thrived with Bolckow, Vaughan and the engine of steam
> Now they have gone and you're alone again

Chris Rea, another singer and songwriter from Middlesbrough, also captured the deep gloom on Teesside during this period in his 1985 song 'Steel River'. Chris's father ran an Italian-style coffee and ice cream parlour on Linthorpe Road in the centre of Middlesbrough. When I was a boy, a trip to Rea's – in my parent's case for coffee, and in mine for ice cream – was a rare treat, long before today's coffee-shop culture made coffee out an everyday and unremarkable event. When he was a boy, Chris used to work weekends in his father's shop, so I guess he may have served me ice cream. In any case, he went on to become a talented musician, and another enduring Middlesbrough export. In 'Steel River', written at about the same time that the first salmon ventured back into the Tees, Chris laments that a few salmon are small compensation for rampant unemployment and poverty.

I'm with Vin Garbutt in rejoicing in the idea that life is returning to the Steel River, but Chris Rea's song also resonates strongly with me – my joy is tempered when I see the difficulties faced by those living in an area that still has one of the highest unemployment rates in the whole of the UK. Between 1976 and 1984, the steel industry shed two-thirds of its workforce – 35,000 people – and between just 1982 and 1987, shipbuilding declined from 18,000 jobs to zero. Clearly, we need a new economic model – one that doesn't set the welfare of working people against the welfare of their environments but instead targets both social wellbeing and the health of the natural world; after all, it's now widely recognised that these two aspects of life are inextricably linked. Thankfully, as we'll discover in our journey upstream, such models are emerging from a new generation of economists who recognise this truth and are prepared to think outside the box.

Middlesbrough is no stranger to devastating unemployment as the current economic models, which have their roots in the 1850s and which developed further in the 1950s to fuel the Great Acceleration, inevitably create cycles of boom and bust. In the boom times, life is tough enough for workers, while those at the top rake it in. However, the wealthy walk away from a bust with pockets bulging, the working people with nothing.

Even during the height of industrial expansion there were times when many found themselves without work. In the second half of the 1870s, unemployment was so severe that a workhouse was built in Middlesbrough to house and feed homeless workers. Able-bodied men worked, perhaps breaking up slag or on some other menial task, in exchange for food and a bed. The workhouse also housed children as well as the elderly. Often, at the end of a life of hard and filthy work, the workhouse was all a labourer could look forward to. My great-great-grandfather, Isaac Brunskill, from the part of my family that originated further up the Tees in Middleton-in-Teesdale, died in the Middlesbrough workhouse in 1920, after spending a large part of his life labouring in the ironworks.

The more recent downturn in industry along the Tees may have helped salmon, sea trout and eels, among others, to recolonise the river, but these fish still don't have it all their own way. Just a few years after those million salmon were released upstream, a tidal barrage was built across the Tees at Stockton. Opened in 1995 by Prince Philip, this was certainly an impressive development, with a world-class white-water course built through a landscaped park adjoining the barrage and feeding off the head of water that built up behind the dam. But the barrage effectively halved the length of the estuary. Prior to its existence, the twice-daily pulse of the tides reached thirty miles upstream, and saltwater penetrated for nearly twenty miles. The barrage was built just ten miles from the river-mouth, halting both the rise and fall of the tides and the effects of freshwater mixing with saltwater beyond this point. It also changed the dynamics of the mixing of saltwater and freshwater downstream of the dam and altered patterns of sedimentation. But from a salmon's or eel's perspective, the biggest problem with the barrage was that it was an impenetrable barrier to both upstream and downstream migration.

To allow for upstream passage of salmon and sea trout, a fish pass, known as a Denil pass, was built into the barrage, and later a fish counter was added to monitor the hoped-for continuing recovery of fish stocks. A footpath across the barrage leads past a narrow concrete channel which forms part of the fish pass, a place where I've waited long hours for the occasional glimpse of an athletic fish beating its way upstream. Unfortunately, the fish pass seemed like a bit of an afterthought and the Environment Agency didn't stipulate how effective the pass needed to be. Even after thirty years of operation, the agency is still not satisfied with the performance of the fish pass. In any case, salmon don't like using fish passes. Radio-tagged salmon have been shown to approach such barriers, then turn back downstream before approaching the barrier again. This behaviour wastes valuable energy, a critical resource on their upstream journey.

Many salmon and sea trout avoid the fish pass completely and opt for leaping over the flood gates themselves when water levels are high enough, although not all make it. A walk over the footbridge across the barrage sometimes reveals scattered carcases, bleached bones that have welded themselves to the ironwork. Some fish choose to bypass the barrage altogether by swimming up the white-water course when it's flowing, which, to fish senses, must appear to be a more natural obstacle as they struggle on to spawning grounds. The white-water course also includes four enormous 'Archimedes screws' which lift water to feed the course when the head of water behind the barrage is low and, more recently, an additional fish pass has been installed between the screws.

The fish counter on the original pass can't distinguish sea trout from salmon but has recorded good numbers of fish moving upstream, although a high count of nearly 1,700 in 2012 was exceptional and probably due to high flows and water levels in the river following a very wet summer. Even in natural rivers, fish runs vary according to the state of the river and, in most years, only around 500 fish are counted at the Tees Barrage. This, of course, is an underestimate of actual numbers, since many more fish avoid the pass and its counter altogether. But the Tees is still lagging behind its two North-East rivals, the Tyne and the Wear. Some 12,000 fish are counted annually in the Wear and 20,000 in the Tyne. Although both the Tyne and the Wear still abounded in salmon when those in the Tees were fast disappearing, these rivers, too, eventually suffered catastrophic declines of fish owing to industrial and sewage pollution. However, effective clean-up campaigns along with declines in industry along both these rivers have now turned the Tyne and the Wear into the best and second-best salmon rivers in England. Salmon recovery in the Tees is at least twenty years behind the Tyne and a tidal barrage certainly hasn't helped.

The problems facing Atlantic salmon are not confined to the Tees. Despite their success in formerly dead rivers, these fish are still in serious decline right around the North Atlantic because they face a barrage of other problems. One is the rapid rise in the number of salmon farms, within which salmon are crammed at unnatural densities. In these crowded conditions, fish lice reach plague proportions. Just ten of these parasites are enough to kill any wild smolt that swims past the cage. It's also possible that other diseases suffered by farmed fish could infect wild salmon. In addition, fish inevitably escape from the farms and breed with wild salmon, diluting gene pools that have become finely tuned to the varied conditions in the rivers in which different populations breed.*

Even electromagnetic fields from cables carrying power from offshore wind farms could be a problem, interfering with the salmon's ability to detect the Earth's magnetic field. Additionally, salmon still face lethal pollutants in their migrations through the North Sea. Even at the start of the twenty-first century, the Tees was one of the main sources for polybrominated diphenyl ethers flowing into the North Sea. These chemicals accumulate to toxic concentrations in higher levels of the marine food chain, including the predatory salmon. Finally, a recent study suggests that climate change, by altering the distribution of the

---

* It's estimated that there are some 2,000 different genetic forms of salmon around the North Atlantic basin.

59

salmon's prey, is proving more disastrous than all these other problems combined.

All of this means that today, fewer than 5 per cent of smolts return to rivers as adult fish. Just two decades ago, that figure was 25 per cent. Additionally, returning salmon are older (lack of food means they must feed at sea for longer) and in poorer condition than in previous decades. Catches of salmon (which now are catch-and-release) in UK rivers have dropped to low levels, with just 50,000 reported in 2018, compared to 600,000–800,000 caught annually up to the 1960s. Such figures have led many scientists to conclude that Atlantic salmon could be extinct as breeding fish in British waters in just a decade, a stark illustration that we need to address local, national and international problems with equal urgency.

Eels are also in big trouble across most of their range. In recent decades, their population overall has declined by 95 per cent and conservationists still don't know exactly why. The number of glass eels struggling upriver varies a lot from year to year, probably due to variable winds and currents in the Atlantic as they make their way back as larvae from the Sargasso Sea. In a good year, with currents at their back, many more larval eels survive the 3,000-mile swim. But that arduous journey across a predator-infested ocean can't account for their recent decline, since they've been making this trek for many millennia. Perhaps it's because we're eating more eels these days, and their unique life cycle means that, unlike salmon, they can't be farmed. Instead, they must be ranched, a process that involves scooping up large numbers of young eels as they return to rivers and rearing them to eating size in tanks. This process depends, however, on collecting eels from the wild. Despite all the problems for wild salmon created by densely stocked salmon farms, they do at least take fishing pressure off wild stocks. The same isn't true for eel ranching. Additionally, on the Tees, as for salmon, the barrage is a barrier both to eels migrating out to sea and to elvers making their way back upstream.

It's not just the physical obstruction to their journey that creates problems for migratory fish. Those fish that cross the barrage in either direction face a sudden change in salinity, rather than the gradual change from saltwater to freshwater or freshwater to saltwater that they would have encountered in the past. This causes additional stress to both smolts and adult salmon, and as adults returning from distant Greenland and Icelandic waters, they need every ounce of their reserves to make it upstream and still have enough energy left to spawn. On top of that, salmon and sea trout pausing downstream of the barrage, working out a way forward, also face a threat from wily predators – seals.

Teesmouth's seals have learned that the downstream side of the barrage is a great place to fish and a couple of seals squabbling over a large salmon is sure to draw an enthralled crowd – real live nature in action. But it's not a sight to thrill those that are funding the recovery of salmon, or salmon fishermen. The concrete buttresses leading up to the fish pass have been covered in iron spikes to discourage the seals from hauling out to digest their meal, although now that seals have learned that there's easy fishing here, they'll be hard to discourage. Both of our two native species,* grey seals and harbour seals, swim upstream as far as the barrage, although it's the grey seals that seem to be eating most of the salmon and sea trout. Numbers of grey seals at the barrage are currently rising each year as word spreads of the easy fishing. The bigger grey seals bully harbour seals and although numbers of the latter are slowly building across the whole estuary, fewer now visit the barrage as more and more grey seals discover the opportunities here.

But, like the salmon, the seals, too, are important creatures

---

* Teesmouth has had a few more unusual seal visitors – vagrants from much further afield. In January 1999, a bearded seal turned up in Hartlepool Docks on the north side of the estuary. This species is normally found in the high Arctic. In 2004, a hooded seal popped up in the Tees Estuary. The normal haunts of this species are the icy waters off Canada, Greenland and Iceland.

– returning to the estuary from which they had disappeared for many decades. Like salmon, they're monitored closely and have their own conservation plans. So, the seals at the barrage give conservationists a bit of a headache. A conflict of interest arises because generally we view nature conservation as if primarily for human benefit. We strive to conserve certain species over others to preserve a human-centred view of nature – but which humans? For example, those managing fish stocks view rare birds like goosanders (fish-eating ducks which we'll meet higher up the Tees) as pests – simply for doing what they've been doing for a lot longer than there have been fishermen to complain. On the other hand, they're a high-priority conservation target for ornithologists. An anthropocentric view of nature inevitably leads to such conflicts of interest. Aldo Leopold, an American ecologist and philosopher of conservation, writing in the first half of the twentieth century, expressed this problem far more eloquently when he said, 'Harmony with land is like harmony with a friend; you cannot cherish his right hand and chop off his left. That is to say, you cannot love game and hate predators... The land is one organism.'*

A more recent trend in conservation is to let nature take over the job it's been doing since life arose three and a half billion years ago, and at which it has had a lot of practice. These new 'non-management' ideas fall under the broad heading of 'rewilding', an approach that isn't suited to all environmental contexts and whose results are not always what conventional conservation either expects or wants, but which is likely to produce more robust ecosystems. Rewilding is a new way of seeing nature as not bent to the will of humanity but free to go its own way. It's a much-needed change of perspective and an important element in a new worldview that must emerge if we're to survive in the Anthropocene.

At most, I only ever see a few seals hanging around the

---

* Aldo Leopold. *Round River*. Oxford University Press. 1993.

barrage, but further down the estuary numbers have built steadily over the past few decades. Early in the nineteenth century, before industrial expansion, there were probably around 1,000 seals in and around the estuary. They haven't yet climbed back to their original population, but in 2016 over 180 seals were counted at one haul-out. Around two-thirds were harbour seals which now once again breed in the estuary, where around twenty pups are born each season.

Harbour seals drop their pups on the sandbanks at low tide, so the pups must be able to swim almost immediately after birth. Seal pups are born with a white covering of hair, called lanugo (even human babies occasionally have a fine hairy coat, also called lanugo, when newborn). Most harbour seal pups moult their lanugo before birth and any that don't moult before they're born do so as soon as they hit the wet mud, making them ready to take to the water as soon as the tide comes in. Grey seal pups keep their lanugo for longer. It takes them three or four weeks to grow a suitable pelage for swimming, so grey seals must drop their pups on rocky islands, above the reach of the waves, or in dunes beyond the range of the tides.

For this reason, grey seals don't breed in the Tees Estuary, but they visit for the fishing from breeding colonies to the north and south of the Tees. To the north, there's a big colony on the Farne Islands, breeding on rocks around the outer islands. Since my first visit to the Farnes half a century ago, I've returned many times to film the wildlife spectacles here. However, no matter how often I visit, I'm always beguiled by these ever-inquisitive seals. Around our boat, first one head, with its distinctly Roman nose, then another pops up from the silky-grey water, large, dark eyes filled with curiosity. Soon we're surrounded by a dozen or more seals, all watching with undisguised disdain as we fumble with our cumbersome underwater cameras. Once we enter their realm, though, they really put us in our place. With barely a twitch of a flipper, they streak through the water, streaming air bubbles from their fur as they twist and turn with ethereal grace around our clumsy camera crew. It's a game for them and more seals join in until our air supply runs out and we have to clamber back onto the boat.

On land, though, it's a different story. To the south of the Farnes, there's another breeding colony at Donna Nook on the Lincolnshire coast. Here, the seals haul up into the dunes, although on this flat stretch of coastline, where the tide retreats for miles, it's an enormous effort for creatures so graceful in the water but so cumbersome on land to get clear of the water. The colony, on a Lincolnshire Wildlife Trust Reserve, is stretched out along the shore, beginning just a few yards from a car park. So, the Trust ropes off the area occupied by the seals for protection, both theirs and ours – male grey seals are extremely large animals with very sharp teeth. But in late autumn, the pupping season, it's still possible to get very close to the young seals, to peer into their large eyes – dark, liquid pools set in snow-white fur.

Most of my early memories of the Tees Estuary seem to be of bitter winter days, bent double against a wind whipping over open sand- and mudflats. But this is when these intertidal mudflats really come alive. They draw large numbers of wintering waders and waterfowl – knot, dunlin, redshank, grey plovers, bar-tailed godwits and shelduck, among others. Around 25,000 birds winter here, even though pollution from growing industrialisation has had major impacts on the estuary. Many of these birds feed on creatures living in the mud and luckily, since the intertidal flats are washed twice daily by cleaner seawater, life was able to cling on in mud and sand even during the worst days of pollution. Even so, the estuary birds still felt the effects of being surrounded by heavy industry.

Larger creatures, like ragworms and shellfish such as tellins and cockles, as well as mud snails (*Hydrobia*), live at much lower densities in the Tees than in unpolluted estuaries. However, there may still be a few hundred ragworms per square yard and tens of thousands of mud snails. Some of the tiniest worms (the so-called meiofauna) are more resistant to pollution and exist in unimaginable numbers in the mud. About half a dozen kinds of worm live at densities of perhaps 1 to 2 million in every square yard of mud. So even in the days when the river upstream was almost dead, there were still birds to be seen on and around the mudflats of the estuary, often in considerable numbers. In the early 1970s, for example, I watched wintering flocks of knot numbering up to 70,000 birds, a spectacular sight as they swirled over the mudflats, like smoke blown from the nearby industrial chimneys.

Different birds target different prey and they feed in different ways and in different places. In this way, they divide up the resources of the estuary, allowing all these species to coexist. Grey plovers, bar-tailed godwits and curlews all seek out large ragworms, but curlews and bar-tailed godwits don't generally feed in the same areas, and thus avoid competition. Some grey plovers wander widely over the mudflats in search of prey, while others establish feeding territories, which they defend against other plovers. These territorial birds may have an advantage over the nomadic ones, since they seem to survive better during spells of bad winter weather. Grey plovers also feed on mud snails, as do knots elsewhere, although on the Tees, knots prefer small mussels. On other estuaries, where numbers are higher, mud snails are also a favourite of shelducks but, on the Tees, shelducks feed mainly on the huge numbers of tiny worms, as do the flocks of dunlin.

It's easy to lose many hours watching all these comings and goings, wondering why a small group of dunlins choose one particular patch while a redshank chooses another of seemingly identical mud, and marvelling at how those bills can detect prey buried in the mud, then grab it, all in the blink of an eye as the birds probe like animated sewing machines. But they're obviously very successful. Between them, these birds have a huge impact on the larger creatures in the mud so that, by the end of the winter, nearly 90 per cent of ragworms, shellfish and mud snails have fallen prey to probing beaks.

From the hide which overlooks Seal Sands, all these little dramas are played out every time the tide drops to reveal the mud. There was no one here to carry out bird counts in the days before heavy industry came to the Tees, so we can only guess what the original wintering flocks of birds would have looked like. But even before the big clean-up, there were still tens of thousands of knots and dunlins, a few thousand shelducks and redshanks, hundreds of curlews and bar-tailed godwits and a few hundred ringed and grey plovers.

The efforts to clean up the river also helped the fauna of mud and sand. Marine creatures that are less pollution-tolerant are once more moving back into the estuary and seaweed is spreading back up the river. Right up to the barrage, exposed rocky banks are now covered in slippery mats of *Fucus* seaweeds. All this suggests that feeding conditions should be improving for the birds, but any benefits this may have brought have been offset by an even greater problem than pollution. One way to reach the Seal Sands hide used to be the mile or so walk along a track appropriately called the Long Drag. It's now flanked by oil refineries and other new industries so is sometimes not accessible along its whole length. However, up until the 1970s, this route didn't exist at all – or it would have been an impossible wade through deep, sticky mud. All this area was once part of Seal Sands.

In the first half of the nineteenth century, there were over 6,000 acres of intertidal sand- and mudflats around the Tees Estuary. Reclamation of the mudflats began shortly after the North Gare breakwater was completed and, by 1900, nearly 2,500 acres of the estuary had been reclaimed. From the 1930s onwards, Seal Sands was dredged to keep river channels open, and the dumped mud and sand raised the surrounding flats by up to twelve feet, creating space for further industrial expansion. By 1969, under 1,000 acres of mud- and sandflats remained. Then, in 1973, a large part of the rest was reclaimed for oil terminals to service the Ekofisk oil field that was being developed out in the North Sea. That left just 370 acres for the birds. Nearly 95 per cent of the original mudflats had disappeared.

For birds largely dependent on these mud- and sandflats, the loss of critical feeding habitat was an even more serious threat than pollution. The final stage of reclamation was scheduled with only a short lead time, so there was little time to record bird numbers and behaviour and draw up a baseline against which to judge the impacts of the loss of the mudflats. Nevertheless, researchers from Durham University, led by Peter Evans, responded quickly and began an intensive study of Teesmouth's wintering waders and wildfowl. Although these studies highlighted the complicated changes that followed the reclamation of two-thirds of what remained of Seal Sands in the 1970s, they've also given us a lot of detailed information on the natural history of these birds on the Tees Estuary.

Those areas earmarked for reclamation were first surrounded by high walls of waste slag, a plentiful building material on Teesside at the height of iron- and steelmaking. These walls were porous, so they still allowed the tide to flow in, although it was delayed with respect to the unenclosed parts of Seal Sands, as it took time for water to percolate through the slag. Before the final stages of reclamation, this may have even benefitted some birds. For redshanks and dunlins, the amount of time they can feed between the high tides is critical. They feed for as long as there is mud exposed, so the delayed tide inside the enclosure gave them extra time.

But the good times were short-lived. In the long term, redshanks and dunlins suffered more than other birds as the land was finally reclaimed. Reclamation is easiest around the high-tide mark, where dry land is already exposed for longer periods, but this removes those critical feeding areas that would normally remain available to redshanks and dunlins close to high tide. It cuts potential feeding time for these birds from around twelve hours to eight hours, which gives them a serious problem. Other species, like shelduck and curlew, only need to feed for around five to six hours during each tidal cycle, so are less affected by land reclamation and the resulting reduced feeding times.

Shelducks also benefitted briefly during the later stages of the Seal Sands reclamation. Once the area to be reclaimed had been surrounded by slag walls, mud was pumped in from the river, to raise it above the level of the tides and ultimately create dry land. But, at first, mud leaked out through the porous slag walls, covering a lot of the sandier areas with fine particles, and creating conditions suitable for the micro-worms that are the mainstay of the shelducks here. So, during the early stages of pumping, shelduck numbers actually went up. But, like the extended feeding initially on offer to redshanks and dunlins, the expanded shelduck habitat didn't last long. Soon, the whole area was dry land and, as industry moved in, the birds had to move out.

In the 1970s, as this latest major reclamation of Seal Sands took place, the sheer scale of the loss of available mudflats concerned conservationists. In addition, the Tees Conservancy Act of 1920 allowed the local port authority to reclaim as much of the intertidal flats as it wanted without recourse to any planning agreements. This licence to kill the rest of the estuary expired in 1984, although the port authority at the time argued that the expiration date should be pushed back. It was vital to prevent any such extension and to preserve what remained, along with associated habitats used by the birds. Step forward Angela Cooper, to whom the word redoubtable simply does not do justice. At the time, my uncle, a retired (and pretty tough) senior police officer, was working in security at British Steel and so had many encounters with Angela as she fought to retain Teesmouth's remaining critical habitats. To judge from his many comments to me, this man, who had taken the hardened criminal community in the rougher parts of Middlesbrough in his stride, was thoroughly cowed by this extraordinary woman. The birds of Teesmouth owe her a great debt. With help from other conservationists, she succeeded in protecting enough of the estuary for it to remain an important part of the winter itinerary of waders and wildfowl.

Peter Evans's studies soon revealed that the lives of all these wintering birds are complex enough even without the loss of so much critical habitat. I met Peter a few times, both on Teesmouth and in Bristol, when he came down to take part in BBC radio programmes to which I was also a contributor. So, I had chance to hear the early results of his work first-hand. I also learned that he came originally from Thornbury, just a few miles north of Bristol, and so had migrated in the opposite direction to me. He died in 2001, at the too-early age of sixty-four, even as some of the longer-term results of his group's research were still being collated. But his work was already showing how shorebirds survived the winter by using different estuaries at different times of the season and how Teesmouth fitted into their complicated flight plans.

Since 1977, some 10,000 waders have been caught and ringed as part of this project and some have been colour-marked with harmless dyes to make them easy to identify in the field. Most of our knot (which are more properly called red knot to distinguish them from the larger great knot) come to our shores from Greenland and Canada. At first, many of them head down to the Wash, arriving there between August and October. The Wash is a vast expanse of mudflats occupying an area that, on a map, looks like someone has bitten a chunk out of England between Lincolnshire and Norfolk. The tide here races across miles and miles of open mud at the speed of a slow trot, pushing knots into swirling flocks as it advances.

At high tide, there's no mud left exposed and the knot pile into high-tide roosts to wait for the tide to turn. There's a well-known roost on the RSPB reserve at Snettisham in North Norfolk which is overlooked by a hide. Here, thousands of knots are crammed together in a huddled grey mass, in one of Britain's best winter bird spectacles. At times they're joined by oystercatchers, more flamboyant with their black-and-white plumage and bright-red bills, although the knots are so tightly packed that the oystercatchers must form their own regimented ranks on the flanks of the knot roost.

The mudflats are not visible to these birds since they're roosting behind the sea wall, but all the birds know exactly when the tide begins to retreat, at which time large groups of knots take to the air, swirling briefly over the roost, before heading off to feed on newly exposed mud. They leave the roost in waves, but eventually all are back on the mud, feeding as fast as they can. They need to pile on the calories, since they've come to the Wash to moult, and replacing feathers takes a lot of energy. After moulting, many of these birds then move up to Teesmouth. The same seems to be true of dunlin and sanderling, since birds ringed on the Wash in late summer or early autumn often turn up on Teesmouth later in the winter. Juvenile birds, on the other hand, don't need to moult in the autumn of their first year and these birds probably come straight to Teesmouth from the Arctic tundra on which they hatched.

Knots on Teesmouth may then move up to the Firth of Forth in late winter, before heading back across the North Atlantic to Greenland and Canada. So, it's not just a question of birds displaced by the reclamation of Seal Sands simply moving to some other estuary. In the minds of these birds, Teesmouth is part of a bigger strategy for winter survival. It was obvious that reducing the mudflats by two-thirds would affect the birds, although it wasn't clear exactly how. Birds that feed on big prey, like ragworms, have already eaten most of them over the course of the winter, so with two-thirds of these worms now buried under oil terminals, the food runs out sooner. Would the same number of birds arrive and then leave earlier for pastures new when the food had run out, simply shortening the Teesmouth section of their winter itinerary?

In fact, lower numbers of birds settled on the estuary in autumn and winter. They seem to have adjusted their numbers to match the reduced area of mudflats, and these fewer birds (with the possible exception of dunlins) stayed for their normal duration. But redshank and dunlin numbers fell by more than the equivalent of the land lost, since these birds were affected both by the overall loss of food and by the loss of feeding times around high tide. The birds displaced by the loss of feeding areas may have gone elsewhere, increasing pressure on other estuaries, or they may simply have not survived the winter. It's very likely that the displaced birds were juveniles, which are subordinate to the adults, so the loss of critical winter feeding grounds may jeopardise future generations of these long-lived shorebirds.

The changing fortunes of Teesmouth's birds have been monitored closely for the last five decades. In the last twenty years, although no further large areas of mudflats have been reclaimed, numbers of wintering birds have continued to fluctuate. Some species have increased in number, benefitting from the creation of pools and wet grazing marshes. Others have continued to decline. Numbers of dunlin and bar-tailed godwits have dropped recently, although changes on the Tees Estuary mirror those seen across the whole of the UK, reflecting an increasing tendency for these birds to winter further north, nearer their breeding grounds, as the climate warms. However, in the last few years, knot numbers have fallen to all-time lows, even though the wintering population in the UK as a whole seems stable. Recently, just 700 birds have been recorded at Teesmouth, down from the flocks of 70,000 that swirled over the estuary when I began birdwatching here – an alarming 99-per-cent decline in numbers.

It's not clear what is driving all these continued changes, although a filamentous green alga called *Enteromorpha* has recently spread over large parts of the remaining mudflats, perhaps responding to excess nutrients in the river water. In addition, the estuary has been altered so dramatically that sedimentation patterns have changed and many areas that were formerly mud are now more sandy. Both these changes make it harder for wading birds to feed. And estuary wildlife continues to face new challenges.

Throughout the nineteenth and twentieth centuries, environmental and social concerns never stood in the way of development along the Tees – indeed, these concerns were usually seen as an impediment to 'progress'. Surely a more enlightened approach prevails in the twenty-first century? You might hope so – but in October 2021 and then again in February and September 2022, enormous numbers of dead crustaceans washed up on beaches around Teesmouth, recalling descriptions from a century earlier of slicks of dead fish brought in on every tide. Local fishermen reported a 95-per-cent decline in lobster and crab catches and the effects were felt as far away as Whitby, twenty-five miles down the coast.

In Hartlepool, just to the north of Teesmouth, the prawn catch is down by 90 per cent and half the fishing boats are now up for sale. At the same time, harbour seal pups have been found dead and dying and those that survive are underweight and in poor condition.

Fishermen and environmentalists were quick to draw attention to renewed dredging operations on the Tees to build deep-water facilities for the newly announced freeport (more of which in Chapter 5). Many of the deeper layers of sediment, laid down in the bad old days and laced with lethal pollutants, have remained largely undisturbed for a century or more. So, perhaps the deep dredging has released a deadly cocktail once again. On the other hand, the government's Department of Food and Rural Affairs (Defra) initially laid the blame on toxic algal blooms offshore. Such things do happen, although not during the winter, when one die-off occurred. However, Defra didn't carry out any systematic sampling to confirm its conclusions and the alga at which it pointed the finger is a species that only lives in warm water – the 'stuff of science fiction', according to environmental campaigner George Monbiot.* Likewise, Jim McMahon, shadow environment secretary at the time, also criticised this conclusion in a strongly worded letter to Defra. Meanwhile, teams of independent scientists from the Universities of Hull, York, Durham and Newcastle concluded that pyridines, released by the dredging and dumping of polluted sediments, could be to blame.

The Teesside Freeport was launched in 2021 as a lucrative government flagship project, with the backing of then-Prime Minister Boris Johnson and the Conservative mayor of the Tees Valley, so there was no chance of it being halted while a more thorough survey was carried out. It seems that there are those who, a century on, still see only one option 'between destroying

* George Monbiot. 'The Dead Shellfish Littering Our Beaches Tell You a Lot About Safety and Secrecy in Britain'. *The Guardian*. 6 June 2022.

your manufacturers and maintaining the purity of your rivers'.*
Thérèse Coffey, upon becoming environment secretary in 2022,
agreed to set up an expert review panel, but refused to publish
their terms of reference. It's hard not to compare this with the
similar problems faced by the Fisheries Board in the 1920s and
1930s, when the big industries pressed to use their own experts
with their own terms of reference to carry out surveys of sources of pollution along the river. Indeed, Labour MP Geraint
Davies, a member of the Environment, Food and Rural Affairs
Committee which ordered the review, said: 'The idea that they
should choose their own judge and jury and that panel should
be both anonymous and meeting in secret, with secret terms of
reference, is completely unacceptable.'† Independent scientists
were never interviewed by the committee and the final report
was conveniently inconclusive, even suggesting that the culprit
was an as-yet-unknown pathogen, continuing that great tradition of gaslighting from a century earlier.

A month before, at the COP15‡ gathering in Montreal in
December 2022, Coffey had proudly boasted of the UK's com-
mitment to the Blue Belt, a series of Marine Protected Areas
around the UK's Overseas Territories. Meanwhile, our domestic
waters, despite a growing network of similar Marine Protected
Areas, remain some of the most polluted in Europe. What really
raised an ironic smile, though, is that in 2019, Natural England,
the government agency responsible for environmental steward-
ship, announced that Teesmouth and the adjoining Cleveland
coast would become part of a new expanded Site of Special
Scientific Interest with the aim of ensuring that the wildlife
of the Tees Estuary has a secure future. It will, said Natural

---

* John Sheail. 'A Barrage of Poisonous Water: Inter-War Research on the River
Tees'. *Water Policy* 2(4), 299–312. 2000. 299–312.
† Shanti Das. 'Mass Crab Die-off: Scientists Say "We Weren't Questioned" for
Crucial Report'. *The Guardian*. 15 January 2023.
‡ The 2022 United Nations Biodiversity Conference of the Parties to the UN
Convention on Biological Diversity.

England, make a strong contribution to the Blue Belt of Marine Protected Areas around Britain. 'The Tees Estuary shows how sustainable development can go hand in hand with environmental enhancement, exemplified by the return of harbour seal to the estuary in the 1980s.'* That statement didn't sit easily with the attitude of the government at the time to environmental concerns around one of its flagship projects. In fact, Teesmouth's seals are in big trouble. At the same time as mass die-offs of crustaceans, most pups born on the estuary were severely malnourished and often exhibited mouth-rot disease. Many were in such poor condition that they had to be put down. These problems may or may not be linked to the continuing development of Teesworks, but in any case, no one has yet looked for the underlying causes. So, has nothing changed in the century since industries here had no regard at all for environmental health? Thankfully, it has.

Although industrial giants like British Steel and ICI have now disappeared, a web of smaller industries still surrounds Teesmouth. In 1988, Ken Smith (then environmental information officer at ICI) and Alan Vittery (from what was then the Nature Conservancy Council) realised that one of the biggest issues holding back nature conservation in areas like the Tees Estuary was that neither industry nor conservationists really understood where the other side was coming from. I can testify to this mutual absence of comprehension from the conversations in the 1970s between my uncle and Angela Cooper – although I got the distinct impression that as time wore on, his admiration for her grew.

The result of that 1988 meeting was the Industry Nature Conservation Association (INCA), which serves as a forum for enlightened two-way conversations between industry and conservation bodies. All too often, the ecological value of sites is

---

* Natural England. 'Estuary Wildlife of the River Tees Gets Increased Protection'. Press Release 7 May 2019.

destroyed through ignorance rather than malice. As the INCA partnership has grown, the two sides have been able to foster a surprisingly vibrant natural history in some unlikely places. A lot of these sites are on private industrial land and so they are, *de facto*, well-protected reserves. It's not the only answer to our environmental problems, but it's a very good one with a proven track record. Conservation is sometimes seen as a choice between sparing land (creating nature reserves managed solely for nature) or sharing land (enhancing nature on land used for other purposes). In the small but populous islands that we live on, finding ever-more-effective ways to share land will be a critical part of our new mindset – and Teesmouth points the way.

In contrast, a recent comprehensive global review of the status of 2,500 individually assessed species in the context of either land-sharing or land-sparing concluded that these species fared far better under land-sparing schemes.* Consequently, the reviewer suggested that the best way to rebuild diversity is to intensify farming in certain areas, using the best technology to thus reduce the area of land needed for agriculture, and manage the rest for nature. This is undoubtedly a sound approach in some parts of the planet and may also be part of the solution in some areas of Britain, but my feeling is that we need both approaches to have any hope of restoring nature to our impoverished land.

We live on a very crowded island where 70 per cent of the land is farmed, and farming has done far more extensive damage to the environment than heavy industry. Upriver, as we explore the natural and social history of the Tees Valley, we'll find that a more diverse natural world once existed alongside people, albeit in times when our national population was

---

* Andrew Balmford. 'Concentrating vs. Spreading Our Footprint: How to Meet Humanity's Needs at Least Cost to Nature'. *Journal of Zoology* 315(2), 79–109. 2021.

lower. Even so, groundbreaking land management schemes in the present day are demonstrating that it's perfectly possible to increase the value of farmland for wildlife if we change the way we think about farming. The work of INCA over the last few decades is an even more dramatic demonstration of the power of land-sharing. Nowhere seems more inhospitable to nature than the industrial landscapes that line the lower Tees and its estuary, yet, with a change of thinking, some of these places are now havens for wildlife. But that change in how we view the world needs to go much further.

As we stand at the start of the Anthropocene, in a world perched precariously on the edge of catastrophic change, we need a new paradigm in which development genuinely goes hand in hand with environmental enhancement, as well as with social justice. The story of the lower Tees through the nineteenth and twentieth centuries (and sadly, on occasions, into the twenty-first century) is a stark reminder of the part played by our outdated mindset – of unbridled growth, no matter what – in both environmental degradation and social injustice. However, our journey around Teesmouth also gives us reason to be optimistic. The river is in far better shape than it was when I first began to explore the estuary, even though there is still a long way to go. In addition, thanks to sensitive management, the marshes and pools that lie inland from the mud- and sandflats, now all part of the Teesmouth Site of Special Scientific Interest, are once more flourishing with life, in both summer and winter. In the next two chapters, our journey takes us across these reborn landscapes and across both these seasons as we meet some of the creatures which are thriving in a human world.

3

Waterland

Summer Marshes – Orchids, Dragonflies and Wildfowl

Aye the banks of the Tees I remember right well
When I think of the places I've seen in my time
The call of the curlew, the ferry car bell
Those rose-coloured skies in the evening

GRAEME MILES,
'Banks of the Tees'

One swallow does not make a summer, but one skein
of geese, cleaving the murk of a March thaw, is the spring.

ALDO LEOPOLD,
A Sand County Almanac, 1949

I spent a lot of time on Teesmouth in the late 1960s and 1970s and the only people I met out on the marshes were other birdwatchers or an occasional farmer. I was hardly surprised. You don't come here for the scenery. Industry hems in all the remaining fragments of nature, although that doesn't mean there's nothing to see. A peregrine falcon will use the top of an old smokestack as a lookout as happily as a remote craggy headland, and flocks of waterfowl crowd onto ponds so long as there's food. They don't care about the view. But in 2009, the RSPB opened a new reserve at Saltholme, complete with a £7-million visitor centre. The same year marked a turning point in human history, now half the global population lived in cities. In this new world, places like Saltholme will become ever more vital.

For eleven years, I was a trustee of the Avon Wildlife Trust, eight of them as its vice-chair. Much of the Trust's work centres around simply making people aware of the natural world, whether by inspiring schoolchildren to find and document nature in their own schoolyards or by encouraging visits to its network of nature reserves. This must be central to the work of any conservation organisation, since there can be no appetite among the public to protect something with which they have no empathy, which they haven't learned to cherish. As a wildlife film-maker I like to think that the ever-increasing popularity of wildlife films over the last few decades has played a role in a greater public awareness of nature and of its fragility, but even the best films can't compare with the real thing. You must be able to smell nature, to feel it, to be surrounded by its presence, in order to really appreciate it. So, I take my hat off to the RSPB and its work at Saltholme.

Although it teems with life today, Saltholme was once as much an industrial landscape as the rest of the estuary. The 1860s saw a resurgence of the medieval salt industry when a thick bed of salt was discovered beneath what was once Salt Holme Farm. By 1894, just under 300,000 tons were being produced each year. Later, the farm was taken over by ICI, which dug clay to use in its cement works. When ICI closed its Saltholme operations, the whole place was almost lost completely when it was suggested that the site be used for landfill. Ken Smith and his colleagues in ICI's Ecology and Environment Department (at least this industrial giant had an Ecology and Environment Department!) didn't think that was such a good idea and, luckily for Saltholme and all the people that now visit the reserve, their view prevailed. Wildlife began to return to the pools and marshes of Saltholme.

In 2007, the RSPB signed a ninety-nine-year lease on Saltholme and redeveloped the whole site, removing contaminated topsoil, creating more pools and reedbeds, putting in footpaths leading to public hides, building floating islands for nesting common terns and even erecting a south-facing concrete wall, pockmarked with holes, to give the local sand martins a head start in digging their burrows into the earth bank behind. And, most importantly, there's a shop and café, where you can sit in front of a panoramic window with coffee and cake, watching flocks of terns ferrying fish to camouflaged chicks on the islands, kingfishers fishing, short-eared owls hunting or the acrobatics of the sand martins that have enthusiastically adopted their tenement-block homes and now pursue insects over the reedbeds to feed their own growing young. The people of Teesside, too, have responded to these new opportunities.

The reserve is busy, summer and winter, with people who would never have dreamed of a day out on Teesmouth ten or fifteen years ago. A whole new set of people now appreciate the rich life of the estuary. This, though, should only be the start of their journey into nature. It's said that 8 million people in Britain are members of environmental organisations, although this is probably an overestimate since many, like me, have membership of several. I believe strongly that the roots of recovery from the crises that we face begin with each and every one of us. Yet, the voices speaking up for nature in this country haven't been anywhere near loud enough – a whisper rather than a shout. If we just enjoy our walks in nature followed by a spot of cake and coffee and leave it at that, improvements to our shared natural world will remain slow and halting.

However, since I'm not averse to coffee and cake, or a little comfort, Saltholme seems to be the perfect place to start my summer exploration of the marshes and ponds. In summer, wet grasslands and marshes present a complete contrast to their appearance in winter – now colour is everywhere, flowers and butterflies, dragonflies and damselflies, all in varied hues, or an occasional glimpse of a yellow wagtail, sunshine-bright, as it hunts among the tussocks. Grasslands are speckled with the dark-crimson flowers of grass vetchling. Even the tree sparrows seem decked out in brighter shades of chestnut than in winter.

Sauntering across the meadows, eyes down, I soon spot a scattering of magenta cones – the flower spikes of pyramidal orchids. Smaller and more delicate than the luxuriant spikes of northern marsh orchids that carpet the damp grasslands behind North Gare, the flower spikes of pyramidal orchids are constructed with exquisite precision. Open flowers, like tiny pink butterflies, cluster around the base of a fresh spike, which tapers to a point of unopened dark-magenta buds to create the pyramid that gives these orchids their name. Yet each spike is unique, so each demands closer scrutiny. I'm endlessly fascinated by orchids since there's more than a touch of the exotic about even the modest British species, but more than that, the natural history of orchids is truly bizarre.

They have sex lives as complicated as any soap opera or cheap novel, intriguing enough to have captivated Charles Darwin, who spent many years studying a variety of British species. Most people associate Darwin with his book *On the Origin of Species*, but few realise that the next book he wrote was on orchids. He used the natural history of orchids to flesh out his thinking on the mechanisms of evolution that he'd outlined in *On the Origin of Species* and to answer some of the criticisms of that book. He spent a lot of time contemplating orchids around his home in Kent, and although he didn't have the pleasure of orchid-hunting on the North Tees Marshes, he'd have found plenty to interest him here – a succession of spectacular species throughout spring and summer.

Early marsh orchids, flowering in May or sometimes in late April, are often a delicate flesh-pink colour, although they vary so widely, from a deep mauve to almost white, that each plant I find seems painted in its own unique shade. Later in the year, northern marsh orchids colour the wetter grasslands with their stately spikes of deep-mauve flowers and, in drier areas, the paler flowers of common spotted orchids, each scribbled with deeper pink lines, grow in colourful drifts. In some places, fragrant orchids grow in impressive pink carpets, although each individual flower is always worth a closer look. Peering down the macro lens on my camera at flowers backlit by the sun, they look like they're made of crystallised sugar. Growing from the back of each flower is a long, hollow tube – the spur. Looking closely at this with the light coming through the flower, I can see that the bottom third or so, furthest from the flower, is filled with liquid. This is the plant's nectar, but why is it hidden away at the end of a long, thin tube? It's all part of the orchid's cunning plan.

Plants with easily accessible nectar attract all kinds of insects eager for a sugar fix. The deal is that in return for high-octane flight fuel, the insect will transport some of the plant's pollen to another flower, so carrying out pollination, but there's no guarantee that a newly refuelled insect will visit another flower of the same kind. The only solution is for the plant to produce copious pollen in the hope that at least a few grains will remain stuck on the insect when it finally decides to drop in on the right species again. Along with the costs of creating nectar, producing lots of pollen like this is a waste of energy and if just a few precious grains finally arrive on another flower, only a few seeds can be fertilised. There must be another way.

Fragrant orchids, along with many other orchid species, have found one method. By hiding their nectar at the end of a long tube, it can only be reached by insects with equally long tongues – in this case, butterflies and moths. Since such insects have an exclusive supply of nectar in these long-spurred flowers, which isn't available to the hoi polloi of sugar-craving bugs, it makes sense for them to remain faithful to such flowers. This makes them more efficient pollinators. In this case, both sides win, but many orchids are not quite so fair.

The marsh orchids scattered over Teesmouth's damp grassland don't produce any nectar at all, which seems strange. Why would insects visit flowers which have no nectar reward? This fact is so odd that Charles Darwin simply didn't believe it. Why would insects, as he put it, fall for 'so gigantic an imposture'? But we now know that about half of all orchid species don't bother producing any nectar. Instead, they rely on fooling insects into visiting the flower spike, which saves these orchids the cost of making nectar. In an insect's mind, anything as big and colourful as an orchid flower must surely be advertising nectar. After all, there are plenty of equally flamboyant flowers growing all around the orchids that are doing just that. Darwin may have thought that insects were too smart to fall for this deception, but it clearly works. Remember those burnet moths on the northern marsh orchids growing in the damp grasslands at North Gare? Studies on early marsh orchids (which also produce no nectar) suggest that their primary pollinators are honeybees, although I've seen plenty of other insects visiting the flowers of these orchids – even if, after a while, they always leave disappointed.

Summer or winter, the ponds and lakes scattered across the marshy grasslands of the Tees Estuary serve as a focus for life. I can think of nothing more pleasurable than to spend several hours sitting by the waterside, watching the comings and goings of creatures great and small. In summer, well-vegetated ponds hum with life, both above and below the surface. Most conspicuous, to me at least, are the dragonflies and damselflies. I always notice these insects first, since I spent many years studying them during my days at the University of Bristol. The main subject of my research was the four-spotted chaser and I soon spot several males of this species perched on prominent sticks in the small ponds along the Dragonfly Path at Saltholme. It's like meeting old friends. To say I know these dragonflies intimately is an understatement. I studied them, quite literally, in microscopic detail, right down to the elaborate architecture of their individual cells. In the electron microscope, cell membranes folded into a complex living origami, enveloping the machinery of the cell, are just as beautiful as the insect itself. Of necessity, life is elegant at every level.

Dragonflies are often described as primitive insects, although this gives the wrong impression. They certainly have a long history on our planet since their distant relatives flew through forests of giant horsetails some 350 million years ago, and they do show features that can rightly be described as primitive, in the sense of meaning 'coming first'. For example, the way that the wing muscles power the wings is simpler in dragonflies than in more recently evolved insects. But 'primitive' conjures an image of a creature outmoded, left behind by evolution; for dragonflies, nothing could be further from the truth. Any creature that has survived relatively unchanged from a time when our own distant ancestors were just crawling out of the water shows that dragonflies must be doing something right.

As far as I'm concerned, dragonflies are the ultimate insects. And if I need superlatives to prove my point, I offer just one. The longest migration of any insect is not, as is commonly thought, that of the famous monarch butterfly of North America, nor the recently discovered multi-generational trek of painted lady butterflies from East Africa to Northern Europe, but of a much less well-known insect, a dragonfly called the globe skimmer, which more than lives up to its evocative name. It migrates from India to Africa and back, across the Indian Ocean, dropping in on the Maldives on the way, often in spectacular numbers.

Teesmouth is a long way, in every sense, from the Maldives, but the four-spotted chasers that are basking in the sunshine in front of me could also be long-distance migrants. About every ten years or so, large numbers of these insects migrate across the North Sea, probably after a mass emergence of larvae from ponds and lakes in Scandinavia, Germany or France. Some historical records even describe swarms of these dragonflies filling the air on the scale of locusts – a sight that's definitely on my bucket list.

The more we understand these so-called primitive dragonflies, the more we realise that they're very sophisticated insects. They show behaviour as complex as many vertebrates and so have recently been adopted by behavioural scientists as model animals in which to study the evolution of territorial, mating and migratory behaviour. It's not hard to witness some of this elaborate behaviour and Saltholme is an excellent place to indulge in a spot of dragonfly-watching. Walk down the Dragonfly Path and pass through the dramatic sculpture depicting the life cycle of a dragonfly, created by Middlesbrough artist

Andrew McKeown, and then explore the pools dug especially for these creatures along a recently opened boardwalk to see the real thing.

The four-spotted chasers that I'm watching are now busily engaged in their own brand of territorial behaviour. Dragonflies can be broadly divided into two types based on how they behave when holding territories – they're either chasers and darters, or hawkers. This also happens to follow their formal classification into families. Chasers and darters belong to the family Libellulidae, while the hawkers belong to the family Aeshnidae. Male chasers and darters defend their territories by perching on a prominent stick or rock, a lookout point from which they fly up to investigate anything entering their air space. Hawkers patrol up and down their territory, flying a regular beat, only deviating to check out anything that catches their eye. And dragonfly eyes are huge – as sharp as any in the insect world. So, not much escapes their attention.

Insect eyes are very different from ours. They're made up of tiny individual crystalline lenses, each of which sits on top of a trans-parent cone, lined with light-sensitive cells. Such a unit is called an 'ommatid-ium' and it contributes the equivalent of one pixel to the final image. So, the more ommatidia an insect has, the higher the resolution of the image it can see. But even dragonflies, with around 20,000 ommatidia in each eye, can't see the world with anything like the resolution that we do. Where insect eyes score over ours is in their ability to detect movement.

I try to creep closer to a four-spotted chaser perched close to the bank, but it soon spots my clumsy movements and it's off. If I keep still for a while, I'm much less obvious as a threat and it soon returns to its favourite perch. Very slowly, I get close enough to see it in glorious detail – constantly twisting and turning its head, which is almost all eye, looking this way and that. It's surveying its domain, spotting even the tiniest of insects that flit across its almost-360-degree field of view. A rustle of papery wings and it's gone, only to return a second or two later to the exact same stick, with some hapless fly clamped in its jaws. If the incursion into its territory is by another male, a brief aerial skirmish usually sends the intruder on his way. But if it's a female, I'll witness another peculiarity of dragonfly behaviour.

Unlike most dragonflies in Britain, male and female four-spotted chasers are similar in colour, although, like all male dragonflies, the male four-spotted chaser has a pair of claspers at the tip of its abdomen. A male flies up to investigate anything shaped like a dragonfly. If it's another species, he generally ignores it, although if he's feeling particularly feisty there might be a brief skirmish – enough to make the point about whose patch this is. If it's a male of his own species, he'll fight, but if it's a receptive female, he'll use those claspers to grab hold of her eyes. It's just as well insect eyes are nothing like our own! But even though dragonfly eyes are covered in hard crystalline lenses, it's usually possible to tell if a female has already mated since her eyes will bear the scars of the male's claspers.

Once the male has locked on, the pair fly off in tandem to mate, which involves no small amount of aerial skill; eight wings and two brains must work in close concert to avoid an unseemly crash. Both male and female reproductive systems, like those of other insects, open near the tips of their long abdomens, which makes mating difficult while the male is latched on to the female's head. However, the male has a handy second set of reproductive organs at the base of his abdomen which he now fills with sperm from his primary reproductive organs at the tip of the abdomen, all the while hanging on tightly to his mate. Then all the female needs to do is to swing her abdomen forward to connect with the male's secondary organs. In this position the pair form a wheel shape – and they can still fly, even in this utterly unaerodynamic configuration. A dragonfly Kama Sutra would surely be a lot more interesting than the human one. As if to emphasise this point, damselflies – smaller, thinner relatives of dragonflies – sometimes form a heart shape when they adopt their 'wheel' position. However, romantic it most certainly is not.

The male dragonfly's secondary organs are complex structures, fitted with all manner of fiendish contraptions. Spoon-shaped devices scoop out any sperm already in the female's reproductive tract from previous matings before he replaces it with his own. It's an excellent way of assuring paternity – although he has others. He also guards his chosen female as she begins to lay eggs, which, in the case of four-spotted chasers, she does by hovering over the shallows and dipping the tip of her abdomen onto the surface. If he's vigilant, he can intercept any intruding males before they have a chance to deploy their own sperm scoops and effectively cuckold him.

Nearby, I spot a pair of common darters also busy laying eggs. They go about this in a slightly different way. A male darter remains attached to his partner after mating, which is an even more effective way of preventing any other male from latching on and scooping out his sperm, but this means the pair must coordinate even more complex aerial manoeuvres as they dodge through the stems of reeds to skim low over the surface so the female can drop her eggs. Eventually, the effort of flying in such close formation seems too much for the pair I'm watching and the male separates. However, he stays close to the female, now flying solo, as she finishes laying her eggs – just in case another male happens upon her.

The small ponds at Saltholme are home to ten different kinds of dragonflies although in 2022, two southern migrant hawkers popped in for a brief visit, the first record in the North-East. This species is just beginning to colonise southern Britain, an indication of our changing climate. There are also six species of damselflies living around these ponds. The males of azure and common blue damselflies are both bright blue, whereas the common bluetail is green, with a bright-blue tip to his abdomen – like a bright taillight. Later in the summer, a larger blue-tailed damselfly also makes an appearance on these ponds.

This is the emerald damselfly, which is easily told from common bluetails by its larger size and its characteristic habit of resting with its wings partly open. This is why in America these damselflies are called spreadwings. Other damselflies fold their wings neatly along their abdomens when perched. Damselflies are much less territorial than their dragonfly neighbours, but when it comes to mating, they follow more or less the same pattern. However, a male damselfly grabs his mate by the front of her thorax rather than by her eyes, which seems a bit less brutal, and in most species the male remains firmly attached throughout egg-laying.

Damselflies insert their eggs into plants (as do hawker dragonflies) but seem very fussy about exactly which kinds of plants and where they are growing. The pair of common blues that I'm watching are flying in tandem, alighting in turn on a selection of floating leaves, each of which the female probes with the tip of her abdomen, often for just a few seconds, before she decides it's not up to her exacting standards and moves on. Favoured spots, by whatever esoteric criteria damselflies judge such things, attract large numbers, perhaps dozens of females, who get busy sawing into the plant tissues with their blade-like ovipositors while the males just sit there, clamped firmly onto their mates with bodies held stiffly upright. Where lots of common blue and azure damselflies crowd into small areas, the attached males look like a forest of little blue sticks. They always appear pretty relaxed, with legs and wings neatly folded. A female, on the other hand, has to cope with egg-laying *and* the weight of a cumbersome male attached to her thorax. But occasionally these dead-weight males come in useful, as I discovered later that summer, when I returned to the estuary to photograph emerald damselflies.

I soon find a pair of emerald damselflies busy laying eggs into a reed stem and I've wriggled close enough to them to get some decent photographs – the pair perfectly reflected in a mirror of still water. The female's abdomen is probing just below the waterline, but she's not happy with this as an egg-laying site. So, she starts backing further down the stem until she's completely submerged. Her body traps a layer of air, so she'll be fine underwater for a short time, but when she tries to emerge again, she gets trapped by the water's surface film and now it's very handy having a dry male attached. He rouses from torpor, revs up his wings and hauls the struggling female clear of the water – so she can repeat the whole process once more on the next reed stem.

While engrossed in watching the surprisingly complex lives of these dragons and damsels, I hear a distinctive plop behind me. But when I look, there's nothing there but a few tell-tale ripples. Scanning the edge of the reedbeds through binoculars, I eventually spot a platform of flattened vegetation – a dead giveaway. It must have been a water vole. Since it's a warm day, I settle down to wait. After thirty minutes, I'm rewarded by the little rodent popping up onto the platform, carrying a mouthful of vegetation. It uses the platform to feed and then to groom its fur, to keep it clean and waterproof – an essential survival tactic for this aquatic vole.

There was a time when water voles were probably the commonest small mammal in Britain and were certainly familiar to many generations of children. The friendly and laid-back Ratty in *The Wind in the Willows* is a water vole, and when Kenneth Graham wrote his much-loved novel at the start of the twentieth century, the population of Ratty's kin ran to many millions. But the twentieth century wasn't kind to water voles. The drive for an ever-more-efficient countryside saw numbers plummet. The dredging of rivers destroyed aquatic and bankside vegetation, leaving no food or cover for the voles. Meanwhile, American mink, escapees from fur farms, ate their way through large numbers of these exposed and hungry voles. From being our commonest wild mammal, the water vole became our fastest-declining. In just seven years, between national surveys in the late 1980s and early 1990s, their population crashed by 70 per cent. They all but vanished from the Tees Estuary and for several years the only place where water voles were even occasionally glimpsed was at Saltholme. The losses nationally were so severe that several conservation organisations, including the Wildfowl and Wetlands Trust, the Wildlife Trusts and the RSPB, began urgent habitat restoration along with captive breeding and reintroduction projects.

The general improvement in water quality and river management across the country, including along the once-heavily-polluted Tees, has helped water voles both directly and indirectly – indirectly because otters are also returning to rivers from which they had once disappeared. In the past, it was often assumed that otters had declined owing to competition with the increasing numbers of feral American mink, but in fact the opposite is true. Otters can more than hold their own against mink. The decline of otters was less to do with mink and more to do with pesticide pollution, habitat loss and persecution. Once these factors, particularly pesticide runoff from fields, were addressed, otters began to spread again, and the number of mink fell dramatically. Some water vole enthusiasts fear that otters also eat water voles, but since otters and water voles used to coexist, often in large numbers, up until the middle of the twentieth century, I'm prepared to bet that the water voles prefer their old neighbours.

The creation of more pools and reedbeds around the North Tees Marshes has been further good news for water voles and when one began to make a regular appearance in front of one of the RSPB's hides at Saltholme, wardens encouraged it with apples and carrots, giving visitors a chance to see this once-familiar creature. Careful management of even the smallest tributaries of the Tees has also encouraged voles, even in Middlesbrough's industrial and urban sprawl.

Ormesby Beck is a tiny and insignificant little stream, but one which took on the dimensions of a major tributary of the Amazon in my nine-year-old mind. It was only a short bike ride from my home, and although I was very definitely *never* allowed to go there, because it ran too close to a railway line for my parents' comfort, that never curtailed my explorer's instincts. After all, there were all manner of bugs to be caught. So, I was heartened to read that the Tees Valley Wildlife Trust's recent water vole survey found signs of these rodents returning to even this little stream. However, there's a long way to go, and the future of water voles on our crowded island is still far from certain.

The vole's feeding platform isn't the only structure I see as I scan the edge of the reedbeds on the opposite shore of the lake. Not far away is another, higher mound of vegetation. Perhaps the nest of a coot or moorhen? Then I catch a glimpse of movement behind the mound and to my delight a great crested grebe levers itself awkwardly onto the platform. It's a beautiful bird, with wide chestnut head plumes and a black cap, but awkward in the extreme as it manoeuvres itself clumsily onto its nest. Its legs are set right at the back of its body, perfectly placed for swimming, like a stern-wheel steamboat, as well as to power it in underwater pursuits of fish, but on land it must waddle, body held partly erect in an ungainly posture that's most unbecoming for such an elegant creature.

In no small way, the diversity of habitats and creatures around me – pools and reedbeds, dragonflies and water voles – is down to this bird. That's because the great crested grebe was, in part, responsible for the formation of the RSPB. In the latter half of the nineteenth century, a fashion craze for decorating ladies' hats with feathers, or even whole birds, spurred a bloody and lethal trade on both sides of the Atlantic. In North America, Florida's vast egret colonies were decimated to obtain egret plumes – long feathers as fine as filigree that looked far better adorning breeding egrets than on the headgear of wealthy women. Often, feathers or wings were ripped off still-living birds, which were then left to die beneath their nests, while their chicks starved to death above.

Egret plumes became worth more than their weight in gold, incentive enough for plume-hunters to track down even the most remote colonies. Smaller birds, from terns to hummingbirds, were stuffed and mounted whole on hats. An ornithologist from the American Museum of Natural History in New York once recorded forty species of birds in just two short walks through the streets of Manhattan – not a single one alive. In 1886, the American Ornithological Union estimated that milliners had, to that point, used some 5 million birds in their hats – *5 million* lives sacrificed for a passing frivolous fashion.

In Britain, those magnificent chestnut head plumes of great crested grebes were much sought-after. But the layer of dense, fine feathers covering their bodies, which provides the grebes with insulation and waterproofing, was also valuable. This was marketed as grebe fur and, turned into muffs, provided insulation and waterproofing for the well-to-do instead. Under relentless attack, grebe numbers plummeted. By 1860, there were probably only fifty pairs left in the whole country, and they'd vanished entirely from the Tees Valley.

As in North America, dead birds became big business. Along with grebes, hundreds of thousands of kingfishers, gulls and terns were killed in Britain and a staggering variety of birds were imported from around the Empire. At the height of the trade, a single London dealer imported 40,000 hummingbird feathers, 6,000 bird-of-paradise feathers and 360,000 feathers from other assorted East Indian species. So great was the carnage, at home and abroad, that a few groups of women in both the UK and the USA rallied in support of the birds.

In 1889, Emily Bateson, wife of the botanist Robert Wood Williamson, living near Manchester, formed the Plumage League, whose members refused to wear anything decorated with feathers, and who campaigned vigorously against a slaughter driven solely by a fashion fad. Around the same time, Eliza Phillips formed the Fur and Feather League based at her home in Croydon in South London. Support for both movements grew rapidly, testament to the passion and organisational skills of a group of dedicated women working in a time before they could even vote for their Member of Parliament. In 1891, Bateson and Phillips merged their two organisations into the Society for the Protection of Birds.

Originally, all its members were women, although since the organisation drew on the upper echelons of Victorian society, it had considerable influence right from the start. The Duchess of Portland became the SPB's first president and remained so until her death in 1954. Shortly after the society's formation, even Queen Victoria stepped in by decreeing that the wearing of osprey feathers by certain regiments would henceforth stop. And after just thirteen years, the society received a Royal Charter from Edward VII and became the Royal Society for the Protection of Birds.

The RSPB was born out of passionate protests by deeply concerned members of the public – albeit well-connected ones. It would take even more forceful action from such women before they gained the right to vote and become part of Britain's democracy, in which now all adults have a voice. More than ever, that voice needs to be raised loudly and clearly in defence of the natural world. History has shown us the power of individual action – a black woman in Alabama refusing to give up her bus seat to a white passenger; a young girl standing outside the Swedish Parliament proclaiming a climate emergency. As the influential American anthropologist Margaret Mead observed, 'Never doubt that a small group of thoughtful, committed citizens can change the world. Indeed, it is the only thing that ever has.' We'd all do well to heed that thought.

Back on the RSPB's Saltholme Reserve, the grebes have shown their appreciation to those Victorian activists by returning to the North Tees Marshes to breed in good numbers. Today, there are probably around twenty pairs nesting on the pools and lakes here, many of which were created only recently by the RSPB. That increase in grebe numbers has been reflected nationally. Back in 1931, when the slaughter stopped, the great crested grebe became one of the first birds to have a national census of its population taken. The Great Crested Grebe Enquiry was promoted in *The Times* and attracted more than a thousand volunteers to help.

Since then, similar surveys, which depend on an ever-growing army of volunteers, have been carried out for all our other birds, summer and winter, and have resulted in the publication of several bird atlases that together provide an invaluable picture of the changing fortunes of birds in Britain. We'll be scrutinising the results of some of these detailed censuses later in our journey. The Great Crested Grebe Enquiry found around 1,200 pairs of grebes nesting across the country, a considerably rosier picture than the 50 pairs of just a few decades earlier. A little over a decade later, in 1944, one pair returned to the Tees Valley and attempted to breed at Charlton's Pond near Billingham, but its clutch was stolen by an egg collector. However, since numbers were continuing to rise nationally, it was only a matter of time before these grebes were able once again to call the North Tees Marshes home.

The expansion of the grebe population across the UK mirrored the expansion of our road system, along with other major construction projects in the years following the Second World War. These projects drove the excavation of large numbers of gravel pits across the whole country. Once gravel extraction finished, the pits flooded and created a new network of shallow lakes that were much to the liking of great crested grebes. By 1975, the national population had risen to 3,500 pairs and a few grebes had moved back to the North Tees Marshes, where they managed to raise chicks for the first time in many decades. They have never looked back and now these birds are a familiar sight, even from the café at the RSPB's Saltholme visitor centre.

The great crested grebe could easily have been the symbol of the RSPB – and perhaps, given the history both of this bird and the RSPB, it should have been. Instead, that honour falls to the avocet, a dramatic black-and-white wader with a thin, strangely upturned bill – a bird that's experienced even more dramatic swings in fortune than those of great crested grebe. At the start of the nineteenth century, it nested all around the south-east coast of England, reaching the Humber in the North but by the end of that century, it had been hunted to extinction. However, it recolonised the East Anglian coast during the Second World War when marshes along a few stretches of the

Norfolk and Suffolk coasts were flooded to make an expected German invasion more difficult. Although the Germans never came, the avocets did. The flooded marshes provided the perfect habitat. They began to breed on Havergate Island and Minsmere in Suffolk, which prompted the RSPB to buy these sites and further improve them to encourage avocets, along with other birds.

When I began birdwatching in the early 1970s, there were around 150 pairs of avocets breeding in this country, but the best place to see them was still the coastal marshes of East Anglia, which meant a trip to Minsmere and Havergate Island. I still remember my first sight of these stunning birds, looking far too exotic to be a British species, and they shared their marshes with other birds I'd only ever read about – bearded tits and marsh harriers. Once they'd re-established a foothold in this country and their habitat requirements had been understood, avocets began to spread both as breeding birds and wintering flocks. By the beginning of the twenty-first century, there were close to 900 nesting pairs, and they were much more widely distributed than they ever had been before their disappearance.

They first nested on the North Tees Marshes in 2008, in time for the opening of the RSPB's new development at Saltholme – a fitting tribute – and they're now regular breeders. Their presence endorses all the work that has taken place to improve the North Tees Marshes and with them comes the hope that other newcomers will also find a permanent home here – birds that I saw for the first time when I went looking for avocets – such as bearded tits, and perhaps even black-tailed godwits, which have also recently re-established themselves as breeding birds in Britain. Both these species are seen in winter on the North Tees Marshes. The Humber Estuary, just sixty miles to the south, is now a stronghold for bearded tits, so perhaps it's just a matter of time. In 2022, the first bittern bred at Saltholme, the most northerly breeding record for the UK, which shows what's possible with a bit of effort. So, to my mind, the avocet is an appropriate symbol for the RSPB because it embodies a sense of optimism. Its example shows that, given a chance, wildlife can recover quickly and dramatically – especially if we make nature a more important part of our modern lives.

The future could be a yet-more-vibrant place, but there's still plenty to enjoy in the present. The great crested grebe that I've been watching is now busy arranging and rearranging bits of vegetation on her nest mound. Male and female great crested grebes are identical; both have those spectacular chestnut head plumes in the breeding season, so I've no idea which this is. However, this grebe clearly has a vision in mind of what the perfect nest mound should look like, even if it seems to be having some trouble achieving it in practice. It finally gives up and slips off the nest with barely a ripple. Head low, it steams across the water, and I see that it's headed for another grebe. Intruder or mate? Definitely a mate.

As the birds close in on each other they begin to shake their heads, at first each alternating with the other, with colourful head plumes and black cap erect. Then they find their rhythm and shake their heads in synchrony. As the ballet intensifies, each bird alternates head-shaking with reaching backwards to run a feather through its bill, again in perfect synchrony with its partner. I make a mental note to return in a month or so to see how the pair are faring. In any case, I'm soon distracted by loud cries overhead – a pair of greylag geese, another bird that has returned from the brink of extinction in Britain, albeit in a very different manner from the grebes.

Greylags are our only native breeding geese, although like

great crested grebes they almost vanished, at least as breeding birds. Historically, they bred widely around Britain, including on the vast marshes around the mouth of an older and more pristine Tees, but they had the misfortune of being large and tasty. They were hunted relentlessly, and their marshy feeding and nesting grounds were drained for agriculture or industry. The last known greylags to breed in Yorkshire did so in 1831 and by the end of the nineteenth century they'd disappeared from the rest of England. They survived as native breeding birds only in the far north and west of Scotland and in Orkney, although Britain still saw a huge influx of greylags each winter from more northerly breeding grounds in Iceland.

In the past, for several different films, I've tracked these Icelandic birds back to their breeding grounds, which has only deepened my fondness for greylags. Some nest in remote areas on lakes surrounded by barren lava flows, but others have forsaken this subarctic wilderness for the bright lights of Reykjavik. Tjörnin Lake, in the centre of Iceland's capital city, is home to a large population of greylags as well as whooper swans. Life is so easy in the big city that these birds don't bother with the hard and dangerous journey to Britain in winter and instead stay in Reykjavik all year round. Most of their Icelandic kin, however, spend winter in flocks around the British Isles, where some mix with the remnants of our native breeding population. And now the greylag's story gets complicated.

Beginning in the 1930s, wildfowlers began to release greylags in various places around Britain. Most of these were from eggs taken from the native wild population in north-west Scotland, although perhaps some wintering Icelandic birds were also captured. These newly established populations, like those on Tjörnin Lake in Iceland, remained sedentary and soon increased in number. Although their original marsh and fen habitats had been largely drained, these resourceful geese adapted to local conditions and, like great crested grebes, they benefited from the new network of lakes that were created as worked-out gravel pits were left to flood. Today, their population has grown to around 150,000 birds.

For a long time, this population was regarded as merely feral and treated with far less deference, or legal protection, than the remaining wild population in north-western Scotland, even though it had been largely derived from those Scottish birds. Yet, this 'feral' population is more British than those of the red kites, capercaillies or white-tailed sea eagles that have been reintroduced using foreign birds, after their extinction or near extinction in the UK, and which are now hailed as conservation success stories. Some think that once a breeding population of feral greylags became established, a few wild Icelandic greylags may even have joined them, this being an easier option than trekking back and forth to Iceland every year, although further complicating the picture in Britain. Ornithologists have recently made the sensible decision to treat the whole breeding population of greylags, remnants of the original wild population and reintroduced birds alike, as one entity, with the status of a returned native. The greylag is back as Britain's only native breeding goose and it's clearly back in a big way.

I watch a flock of thirty or so birds on final approach to Saltholme Pool. They twist and turn as they descend past flare stacks and pylons, spilling air from beneath their wings to shorten their glide paths, webbed feet splayed to act as air brakes, until they splash down on the pool in a cacophony of braying honks. Such flocks are a relatively recent sight on the Tees Estuary. Back in 1963 and 1964, forty greylag goslings were released at Wynyard Park, to the north of Stockton-on-Tees, but this group died out in the 1980s. Then, in 1979, five geese were seen at Teesmouth, a vanguard of the expanding population. Today, there are probably well over 1,000 geese in the area and their skeins, strung out across an evening sky, make a dramatic sight, and create an evocative soundtrack to the Tees Estuary.

The greylag skeins are not the only ones to cleave the skies of

Teesmouth, but there's no discussion about the status of these other geese in Britain. Canada geese, as their name suggests, are definitely not native to these shores, although, like the grey-lags, they've undergone a huge population explosion over the second half of the twentieth century. They were first introduced to Britain in 1665 by King Charles II to add to his wildfowl collection in St James's Park in London. Others were introduced across the Channel, to the gardens at Versailles, under King Louis XIV, but they remained scarce in the intervening centuries. It's estimated that in 1950, the national population was still only around 2,000–4,000 individuals, and two Canada geese seen on the North Tees Marshes in 1957 were the first recorded in the twentieth century. Even as late as the 1980s, there was still only a small population in the North-East. Today, the national population has risen to about 82,000 and there seem to be flocks everywhere on the North Tees Marshes.

Breeding Canada geese don't mind sharing lakes with non-breeding birds or even other breeding pairs, which means a lot of geese end up crowding together on favoured lakes. Around the North Tees Marshes, Canada geese in flocks several-dozen-strong graze among sheep or cruise in flotillas across the lakes all summer long. They've learned that they have little to fear from people here, and birds breeding on the Saltholme Reserve often bring young chicks right up to the viewing windows at the visitor centre – little balls of fluffy yellow down who are very unsure as to why they've been given such large, webbed feet. As charming as these sights are, Canada geese have reached such large populations that in some places they're now seen as pests. They're serious competitors for grazing with farm animals, as well as producing prodigious quantities of droppings. Even so, a flock of dozens of birds sweeping in against a summer sunset is a breathtaking sight.

A couple of months later, I'm back on the marshes to see how the season is progressing. It's now August and many birds are busy with well-grown chicks. The fluffy Canada and greylag chicks are now scruffy-looking adolescents, trailing their parents and keeping up an insistent nasal whistle. A steady stream of common terns arrive back from the river-mouth, each with a fish clamped in its bill, to feed chicks that look to me as if they're large enough to fly and catch their own. The flight lines of the terns are well known to the local black-headed gulls, who intercept returning birds. Terns are agile fliers, but burdened with a heavy sand eel, they're less manoeuvrable. Still, they put on a good show, twisting and turning to shake off their pursuers. If they can stay one step ahead of the gull for long enough, the gull gives up – there's no point in burning more energy than it would gain from eating the fish. But sometimes the harassed tern drops its catch and then the gull displays its own agility. In a twisting dive, it grabs the fish in mid-air and swallows it, before a bigger gull plays the same trick on it.

However, I can't see the great crested grebes anywhere. The nest platform is still there but looks neglected and a little derelict after summer storms. Then a grebe emerges from the reeds, and behind it, in close pursuit, two well-grown youngsters. Their heads and necks are decked out in natty black-and-white stripes, very different from their parents, who are beginning to look a bit worn and weather-beaten. When they're very young, grebe chicks ride on their parents' backs, but these chicks are now much too big. Instead, they follow a parent everywhere, constantly begging for food. The adult dives and pops back to the surface a few seconds later with a fish in its beak. Both youngsters instantly spot it and steam towards the adult at full speed. The first to arrive gets the reward and within seconds, the adult dives again and once more returns with a fish. Yet again, it's harried by the chicks as soon as it surfaces. There's no rest for any of the marsh birds at this time of year.

Then, without warning, another bird explodes from the cover of reed stems and hurtles towards the grebe family, a whir of wings and legs across the surface. As the grebes scatter, I realise the bird is another grebe, but not a great crested. It's a little grebe, sometimes called a dabchick, and it's half the size of its bigger cousins. But what it lacks in size, it more than makes up for in feistiness. The great crested grebe family, engrossed in fishing, had clearly drifted too close to the little grebe's own family – and paid the price. The little grebe yodels its victory – a braying, far-carrying call that's much louder than anything produced by a great crested grebe – before it slips back into the forest of reed stems.

Several different kinds of ducks also breed around the estuary. The shelduck, symbol of the Teesmouth Bird Club, is one of the most obvious, in its black, white and chestnut plumage, and one of the most peculiar. In late winter, pairs of shelduck set up feeding territories which will be used later by the female and her chicks. But it may be a long trek for the new family, since a shelduck must also find a suitable hole to nest in and this may be some way from the feeding grounds. They often use abandoned rabbit burrows, and on Teesmouth, shelducks are also happy to nest in holes in the walls of slag that now criss-cross the estuary. In fact, slag walls are just about the only place they do nest and it's doubtful whether any would now breed around the estuary at all if not for this legacy of industry and reclamation. However, the lower Tees Valley doesn't have a big population of breeding shelducks. Around sixty nest in the area with perhaps a couple of dozen pairs around the lower stretches of the river and the estuary.

That said, although shelducks in the past were largely coastal ducks, they've recently been moving inland. On Teesmouth, they began by tracing the valley of Greatham Creek upstream, after which they spread more widely through the Tees Valley. Inland, their nests have been discovered in straw bales or under barns. A surprising variety of ducks nest in holes, although usually in holes in trees. Many of these, like goldeneyes in this country or Carolina wood ducks in North America, also happily adopt nest boxes that have been thoughtfully provided by conservationists. But the burrow-nesting shelducks are fussier. In the past, in an attempt to boost numbers of breeding birds, nest boxes were erected but were largely ignored by Teesmouth's shelducks, although the RSPB is trialling new schemes here, to see if it can find something which meets the shelducks' exacting standards.

Despite their careful choice of nest burrow, not all shelducks are such devoted parents as the grebes. Long before their chicks can fly, many of Britain's shelducks make a summer migration to the Heligoland Bight on the other side of the North Sea, in north-western Germany. There are vast mud- and sandbanks here, a long way offshore, where the shelducks assemble to moult their flight feathers. This process leaves the shelducks flightless until the new flight feathers have grown a month or so later, but they're safe on the Heligoland Bight's remote sandbars, which are inaccessible to predators. There may be 100,000 shelducks here and many of them will have abandoned broods of young chicks back on their breeding sites, although this is not such a dereliction of duty as it sounds.

A few shelducks stay behind on the breeding grounds to look after all of the chicks. The other birds simply dump their broods in a vast creche under the watchful eye of these few nursery maids. Creches frequently contain between twenty and forty black-and-white chicks, but sometimes the nursery maids really have their work cut out. A single creche of 200 chicks was once recorded on the Wash and creches of around 100 chicks have been reported from a few other areas. Most creches I've seen are much smaller, but even thirty or so shelduck chicks can be quite a handful.

The antics of chicks in a creche are endlessly entertaining, at least to a human observer. Intent on exploring, the chicks slowly drift apart until the furthest ones realise that the adults are now just distant specks – at which point they paddle furiously over the water, back towards their surrogate parents. Their panic spreads throughout the rest of the creche and suddenly there are dozens of chicks skittering as fast as they can over the surface, converging on the long-suffering adults. Then, panic over, they start exploring again as if nothing had happened. It seems like an impossible task to shepherd such large numbers of increasingly adventurous chicks, but it's obviously successful. Ducklings in large creches grow faster than those that remain in family broods.

At the other extreme from the specialised habits of the shelduck is the archetypal duck – the mallard, usually just called the wild duck in North America. Although they nest on the North Tees Marshes, this is also the duck of park lakes, farm ponds and indeed any body of water larger than a puddle. But its familiarity shouldn't blind us to its beauty. The drake has a metallic sheen on his head that in sunlight glows in ever-changing hues of green and purple. In a world where children have less and less contact with nature, feeding these ducks at the park is one of the easiest ways to come face-to-face with natural beauty and, personally, I have a feeling that these ducks were responsible for some of the very first stirrings of my life-long attachment to nature.

One of the easiest places to appreciate the charm of mallards is not on the marshes of the Tees Estuary, but in Middlesbrough itself, in Albert Park, a recently restored Victorian park with a lake full of very friendly ducks, whose laughing quacks can't fail to raise a smile. The park, if not the ducks, was originally the vision of Henry Bolckow, who not only brought iron- and steelmaking to the Tees but was fledgling Middlesbrough's first mayor and Member of Parliament. As Middlesbrough grew, Bolckow, taking a view well ahead of his time, realised how critical green spaces were, as places where people could escape from the hellish world of furnaces and rolling mills to breathe cleaner air and stroll among trees and shrubs. The park was opened in 1868 by Prince Arthur of Connaught, one of Queen Victoria's children, and named for his father, Prince Albert.

According to my parents, I learned to walk here, out in the

(relatively) fresh air. At the time we lived with my grandparents in York Road, just a short walk from Albert Park, and taking me to see the ducks was apparently a guaranteed way to keep me quiet. Of course, I can't remember that far back, but some of my earliest childhood memories are certainly of this park, of its secret hideaways in dense shrubberies, of treasures like burnished conkers and of a lake full of ducks, all eager for my bag of stale bread. I also fished the lake for sticklebacks, which I kept in jam jars to observe in more detail. Once, when the lake had to be drained for maintenance, I was amazed to discover that the lakebed was carpeted in huge freshwater mussels. I had no idea these creatures even existed until I fished their impressive shells out of the stinking mud. So, thank you, Henry Bolckow, for your early realisation of the vital importance of contact with nature. I learned a lot from that park, including one of my earliest ornithological discoveries – that only female mallards quack. The males have a quiet, slightly raspy 'yeeb' sound. I later found out that this is also true of many other ducks – the females are the noisy ones.

Mallards are known as 'dabbling ducks', since they feed from the surface, upending to reach down into the water, but there are also 'diving ducks'. The commonest breeding diving duck in Britain is the tufted duck, a familiar sight today on many lakes and reservoirs. This is another bird whose population expanded during the twentieth century. Even at the start of that century, T. H. Nelson noted in *The Birds of Yorkshire* that it was already increasing locally as a nesting bird. This early growth may have been helped by the accidental introduction of an alien species of freshwater mussel, the zebra mussel, which turned up in London Docks in 1824. There are still fears that zebra mussels will oust native freshwater mussels, such as those I found in my youth in the lake in Albert Park, but at least they provide food for tufties. Like greylags and grebes, tufted ducks also benefitted from the spread of flooded gravel pits around the middle of the century. On Teesmouth, there was a

staggering increase in numbers in the last quarter of the twentieth century. In 1976, just two broods and a total of thirteen
ducklings were recorded. By the end of the century, that had
risen to 200 ducklings in 33 broods.

Several other kinds of ducks, both dabblers and divers, nest
on the North Tees Marshes, but in very low and variable numbers. However, the creation of new habitats and the careful
management of existing ones means that the populations of
these scarcer ducks are probably secure. Gadwall are closely
related to mallards, but less raucous and the males less colourful. Even so, a male gadwall, formally dressed in grey and black,
is still a smart bird, although a lot harder to find than the ubiquitous mallards and tufted ducks.

It's thought that almost all the gadwall that now breed in
the UK derive from reintroduced birds – or more accurately
from wild wintering birds that were prevented from migrating to their breeding grounds to the north and east of the
British Isles. Like many ducks, wintering flocks of gadwall
were trapped in duck decoys, which were long, funnel-like
netted enclosures covering a pond, open at the wide end but
gradually tapered to a small blind-ended net trap called the
pipe. In the nineteenth and early twentieth century, commercial wildfowlers harvested large numbers of ducks and geese
from Teesmouth, both by shooting and by trapping them in
decoys, an indication of the numbers of both wintering and
breeding birds around the estuary. According to Nelson, who
regularly hunted wildfowl around Teesmouth, a decoy here
once caught 500 ducks in one go.

The trick to such a large 'bag' from a decoy is to persuade
the ducks to swim into the open end and then on into the business end of the trap. For this, the duck trappers enlisted the
help of a dog. If a duck spots a predator like a fox, it will keep
its distance but follow it to keep a wary eye on it. Decoy dogs
have been bred to look like foxes in colour and size, and they
run along the outside edge of the decoy, drawing the ducks after

them. To make the ducks even more curious to follow, a line of small wicker blinds is often arranged down the side of the decoy and the dog is trained to weave in and out of these, so it seems to the ducks that the predator keeps appearing then disappearing. This just makes them all the more eager to track its progress. Decoys and dogs are still used to trap ducks, although today the ducks are merely ringed and sent on their way. In the past, they ended up in the pot.

In 1850, at a duck decoy in Deringham in Norfolk, a lucky pair of gadwall were merely wing-clipped, so they couldn't fly, and then released to breed. Chicks should learn their traditional migration routes by imprinting on their parents and then following them on the journey. The chicks born to the wing-clipped Norfolk gadwall, on the other hand, simply learned to stay put with their flightless parents, and soon a local population of breeding birds arose. Breeding populations elsewhere are the result of further introductions of captive-bred birds, taking place from the 1950s to the late 1970s, particularly in Kent, Cumbria and Leicestershire.

It took a while for gadwall to find the North Tees Marshes but, in 1994, they eventually began to breed here, and now, in most years, there are close to thirty pairs. They need ponds that are rich in nutrients, with a lush growth of vegetation, since, even as chicks, they're more vegetarian than other ducks. Most other ducklings include high-protein invertebrates or even tadpoles in their diet to speed their development. The gadwall's lower-calorie diet also means that the chicks must spend longer foraging than other ducklings, and so are more prone to disturbance. Access to the marshes where gadwall nest at Saltholme is now carefully controlled during the breeding season, to the benefit of many other birds as well. However, several hides overlook good gadwall ponds, so it's still possible to admire this retiring duck without disturbing it.

On their breeding lakes, gadwall are quiet and unassuming, going about their business without attracting too much attention. So, it always feels like a privilege to see a pair slipping along channels through the reeds or sitting on an island of collapsed vegetation, preening. The pair I've found are doing just that, and after finishing their feather care, they stretch their wings, revealing bright-white wing patches, the most obvious way to distinguish the brown female gadwall from the very similar female mallard. Shortly, the drake will lose his dapper grey and black and acquire his dull 'eclipse' plumage, which will leave him looking very much like the female. It's midsummer and they're about to moult their flight feathers, rendering them flightless for around a month, and therefore very vulnerable. Shelduck solve this problem by heading off to remote sandbanks in the North Sea, but most other male ducks rely on the camouflage provided by sombre eclipse plumage. By late summer on the marshes, all the ducks look very dowdy, but as soon as their flight feathers have regrown, they'll moult their body feathers again. Then they're ready to head to their winter quarters, a time when, newly moulted, the males will look at their very best. As I leave the marshes at the end of summer and bid farewell to the orchids and dragonflies for another year, I look forward to another visit in early winter, when the waters will be alive with flocks of wildfowl and waders.

4

Eye of the Wind

Winter Marshes – Wildfowl, Waders and Wind

Like winds and sunsets, wild things were taken for granted until
progress began to do away with them. Now we face the question
whether a still higher 'standard of living' is worth its cost in things
natural, wild, and free. For us of the minority, the opportunity to see
geese is more important than television.

ALDO LEOPOLD,
A Sand County Almanac, 1949

E very year, I can't help feeling a tinge of sadness as summer draws to a close. Spring's promise of new life and a fresh start has been fulfilled and now the world begins to hunker down for winter. The vibrant greens of infinitely varied hues that painted spring and early summer are replaced by the dull monotony of leaves worn and battered by a summer of harnessing the power of the sun, their job now almost done. Forests are gloomy, and grasslands are turning scruffy and brown. But sadness doesn't linger. Late summer is also its own season, with its own treasures. It's a time of grasshoppers and bush crickets, now fully grown and singing their raspy songs, each one as distinctive as any bird's song – and darter dragonflies remain on the wing, sometimes right into November, until the first frosts finally end their season. Now, as I look ahead to winter, my mood lifts even more. Insects and flowers will shortly be replaced by the yearly influx of ducks, geese and waders. By then, male ducks have emerged from their dull eclipse plumage and will once more be resplendent in all their finery. Soon the marshes will be crowded with colourful flocks escaping the harsher winters to the north and east of our shores.

Late summer and autumn are times of excitement in the birding world as migrants aiming for final destinations further south call in on their way through. Anything can turn up anywhere and a real rarity is sure to draw big crowds of birders, but even more common migrants are worth seeking out at this time of year. Waders from more northerly climes begin moving south as early as the second half of summer and arrive here still decked in their breeding colours, which are a lot more spectacular – and distinctive – than their muted winter garb. Dunlin sport jet-black bellies, red knot still look red and black-tailed godwits haven't yet lost their rich russet underparts. Some of these waders will winter here on the Tees Estuary, splitting their time between the mudflats of Seal Sands and the muddy edges of marshland pools. Others, like golden plovers, stick to the grassy fields and marshes. But, as autumn becomes winter, all these waders turn as grey as the weather – a contrast to the colourful ducks that flock here in winter.

In fact, this winter visit to the North Tees Marshes couldn't be much greyer. A freezing wind is blowing snow horizontally across the flat landscape, dense enough to hide the signs of industry all around in a blur of grey. Even the Transporter Bridge is just a ghostly shape hovering over the marshes. Once more I'm grateful for the warm haven of the café at Saltholme, a place to fuel up on hot coffee before setting out into the blizzard. And there's plenty to see without ever leaving the luxury of the centre, since there's also a café here for the birds, an array of feeders hanging close to large picture windows overlooking the lake. And today it's busy with customers.

On the ground, a dozen or so stock doves are trundling back and forth, hoovering up seeds spilling from the frenzy taking place on the feeders above. Stock doves must be one of the most unnoticed birds in the country, and those who do notice them often describe them as looking like dull woodpigeons. Not very flattering for what, close-up, is a very beautiful bird. It might lack the woodpigeon's white neck and wing patches, and in dull light it does appear a more subdued grey than its larger and brasher cousin. But the feathers on its neck are iridescent and in bright light shine with an array of colours that are impossible to describe in words. Field guides label the neck patch as bottle green, but that doesn't do it justice. Even on this gloomy day, the feathers catch light spilling out from the window and create ripples of violet, blue and indigo as the birds move their heads to feed.

I've heard it said that the 'stock' of their name came from the fact they were caught or kept for the pot, but it's more likely to be derived from the Old English *stocc*, meaning a stump or tree trunk. Unlike woodpigeons, which build a scruffy nest from what looks like randomly dropped sticks, stock doves nest in holes in trees, although holes in cliffs or even rabbit burrows will sometimes do. They're widely distributed across

the country, but little known by non-birders, perhaps because they haven't moved into gardens to take advantage of feeders in the way that woodpigeons have in the last few decades – at least, not yet. In most places, they remain resolutely rural, even though here on the North Tees Marshes they're more than happy to take advantage of the free food on offer. I now also know of quite a few other places where stock doves take advantage of our generosity, so perhaps this strategy is catching on and will spread more widely. After all, it's only recently that wood pigeons spotted the opportunity created by a dramatic upturn in the popularity of feeding garden birds and became familiar backyard visitors. Watching the stock doves along with the mêlée of small birds vying for spots on the feeders, I can't help thinking about how much we humans are changing the natural world. In the first part of our journey, we could hardly miss the disastrous effects of expanding industry on the life of the estuary. Yet, we also impact nature with kindness and generosity.

A few years ago, I made a film called *Unnatural Selection*, which documents the many surprising ways in which the world we've created is changing the course of evolution. In this case, I don't mean new behaviours – like those of woodpigeons – as animals learn to adapt to changing circumstances, but actual changes in the DNA of these animals – alterations in their genetic code that parallel the way natural selection works to drive evolution. Hence the title of my film. During filming, we travelled to Montreal University to speak with Andrew Hendry. Andrew had made detailed studies of Darwin's finches on the Galapagos Islands off the coast of Ecuador, where these birds are often drawn upon to illustrate a process in evolution called 'adaptive radiation'. In the distant past, a single species of finch, windblown from its home in South America, found itself stranded on the Galapagos Islands – and in a world full of new opportunities. Here, it evolved (or radiated) into many new species, each one taking advantage of a different vacant niche.

Darwin's finches evolved to exploit the many kinds of food they discovered on the Galapagos Islands and to cope with this new diversity, from heavy seeds or small insects to soft fruits or tough plants, evolution crafted a variety of different beak sizes and shapes. Today, evolution hasn't stopped, and Darwin's finches are still busy radiating. Andrew discovered that one species, the unimaginatively named medium ground finch, is in the process of splitting into two, as separate populations adapt to slightly different-sized seeds. There are now two populations of finches with slightly different beak sizes, not enough to be considered different species yet, but well on the way. At least, that's what was happening before we humans also colonised the Galapagos Islands.

Both populations of medium ground finches now feed on scraps and waste around settlements, an abundant source of food that doesn't need specialised beaks. Natural selection has lost its edge and the two populations are now collapsing back into one. It's as if we've thrown evolution into reverse. Watching the hordes of tree sparrows, starlings and goldfinches, along with reed buntings, greenfinches and assorted species of tits gorging themselves on the feeders at Saltholme as a bitter wind ruffles their feathers and snow blows around them, it's clear that abundant resources like this, in an otherwise harsh season of deprivation, must have big effects on the ecology and behaviour of these birds. In recent years, several studies have been carried out that show the surprising extent of those effects.

Feeding birds in gardens has grown ever more popular over the last few decades and the variety and quality of bird food have increased as the market, and therefore profit margins, have expanded. In the past, birds had to make do with stale bread crusts. My grandmother also used to fry up the rinds of bacon until they were crisp, a fatty treat much appreciated by the local starlings – and by me, surreptitiously scoffing them while I was scattering the pieces for the birds. But today, I offer visitors to my garden a smorgasbord of seeds, protein-enriched fat-balls and dried mealworms. Not surprisingly, my garden is often packed with birds in winter and this same story is repeated across a vast number of other gardens. According to recent surveys, around three-quarters of households in the UK provide some form of food for birds. Even more impressively, between the US, where bird-feeding is now just as popular, and the UK, we supply nearly half a million tons of supplementary food a year. That's enough to feed around 300 million tits (or chickadees, if you're in the States).

We don't really know a lot about the effects of such large amounts of extra food but, given the influence of much smaller quantities of food on the evolution of Darwin's finches in the Galapagos, we must be having a substantial impact. It's thought, for example, that a large and easy supply of winter food is at least partly responsible for changes in the migration patterns of European blackcaps. In the past, these little warblers all left the UK to spend the winter around the Mediterranean – as our breeding birds still do. But blackcaps from Central Europe now don't bother with a long and dangerous journey south. Instead, they make the much shorter journey to our shores. Once here, they stake a claim over garden feeders with a feistiness well beyond their size. They bully most other garden birds, including robins, normally the most aggressive garden visitors, as they defend their winter territories. Blackcaps, however, have no choice but to behave like louts.

Most winter visitors to gardens can find wild food in the countryside and in mild weather they even prefer their natural diet, only resorting to gardens when the going gets really tough. But blackcaps are not adapted to spend the winter this far north and are entirely dependent on garden feeders for their survival. So, they must defend their vital supplies with all the force they can muster. It's life or death for them. The same seems to be true of great tits overwintering in the much harsher conditions of Finland. They couldn't survive the winter there without feeders. This has led some scientists to see bird feeders as 'ecological traps', drawing birds into areas where they can only survive with our help.

Winter is a critical time for all small birds everywhere, so the vast amounts of food we provide makes a massive difference – and it's not just blackcaps and great tits that are taking advantage. In North America, American goldfinches and northern cardinals are extending their range northwards, at least partly in response to the copious supplies of winter food, although climate change is probably another factor. In Britain, winter food is likewise a factor in the northward expansion of nuthatches.

Many people stop feeding in spring, but the effects of winter feeding linger on. Birds that have had the benefits of generous human neighbours over winter are more successful breeders. Often, they lay eggs earlier and, in those species which have been looked at in detail, these early broods survive better. However, some species, such as most of the tits, need to time their broods to coincide with the peak of insect abundance in the spring, so early laying may affect this vital synchronisation. A similar uncoupling of the cycles of birds and their prey caused by our changing climate is already a big worry and winter feeding may only add to the problem, at least for some species.

Supplementary food may even affect sex ratios in bird populations. Sex determination in birds is the other way round from ours. In mammals, males have two different sex chromosomes – an X and a Y – whereas females have two Xs. In birds, it's the males that have the two similar sex chromosomes and females that have two different ones. Bird sex chromosomes are designated W and Z, so males are ZZ and females ZW. This means that, in theory, a female could choose the sex of her offspring by favouring eggs with either a W or Z chromosome. And that's exactly what they do – in response to the prevailing environmental conditions. In species in which the males are bigger, and therefore need more food to grow, females produce more males when there is plenty of food around. Winter feeding might create the perception of such favourable conditions and cause females to produce more males.

This seems to be what happened to a very rare parrot from New Zealand – the charismatic kakapo. Conservationists began providing supplementary food to boost the survival chances of the handful of remaining birds, but after a few years they noticed that very few females were being hatched. Once supplementary feeding was stopped, the sex ratio returned to normal. I don't know of any studies that have looked at such effects in our familiar garden birds, though I'm sure it would be revealing.

Feeding might also affect the relative abundance of different species since not all birds visit feeders, and those that do may fare better than those that don't. However, this factor is changing as more species learn to take advantage of our generosity. In parallel with the spread of feeding and of higher-quality foods, some species have changed their behaviour radically. In the 1970s, only sixteen kinds of birds were regular visitors to garden feeders. Now there are twenty-three, with some having been added to the list in just the last few decades. If you could travel back in time, the main difference you would notice in your garden would be the scarcity of long-tailed tits, woodpigeons, great spotted woodpeckers, nuthatches and gold-finches. It's almost impossible to believe, as I sit over the last of my coffee watching dozens of goldfinches flying around the feeders, that this is a recent phenomenon.

Certainly, in my youth, my classic image of goldfinches was of autumn flocks flitting between teasel or thistle heads, twittering their bouncy calls as they used their tweezer bills to extricate the small seeds. Since the RSPB has encouraged the growth of these plants on the reserve, you can still see such sights around Saltholme, but goldfinches are not dumb. In the past, nyjer seed was used to tempt goldfinches into visiting garden feeders on the basis that these tiny, hard seeds are similar to their natural food items. But in my garden, goldfinches now ignore these seeds in favour of sunflower hearts, which are easier to eat and more nutritious. Many of my bird-feeding friends are finding the same.

This perhaps shouldn't be a big surprise. Goldfinches have proved themselves to be smart in other ways. In the past, they were often kept as pets, for both their colours and their song. But an additional entertainment was to dangle a piece of food on a string below the bird's perch. The goldfinch would then haul up the food, carefully standing on the string after each pull to avoid letting the prize slip back down. Today, this trick is used by scientists as a test for intelligence in birds. Under more careful observation, it was discovered that about a quarter of goldfinches spotted the solution to this problem with no help. Another quarter picked up the trick by watching the smart ones, while about half of them, rather unkindly called 'duffers' by the scientists, never got the hang of string-pulling. Interestingly, siskins – birds we'll meet in abundance higher up the Tees – were even better at this. Over half of them worked out what to do without help. Perhaps the different feeding style of siskins in the wild, foraging acrobatically among fine twigs, makes them more adept at solving the string-pulling problem.

Experiments like these have forced scientists to reappraise

the intelligence of birds. In the past, they were seen simply as creatures driven by blind instinct; today, we know that they have rich inner lives and sharp minds. Indeed, new studies have shown that bumblebees are also capable of solving the string-pull test and that many other kinds of insects possess a degree of consciousness unsuspected just a few years ago. This isn't just of intellectual interest – it should inform the new relationship that we must all develop with the natural world if our human world is to survive the Anthropocene. We share the planet with many more sentient beings than we might imagine.

Feeding stations undoubtedly affect the behaviour and ecology of many kinds of birds, and perhaps even their evolution, but by encouraging them into gardens we also create a valuable opportunity to get closer to nature. It's vital that people experience nature first-hand whenever and wherever they can and in so doing come eye-to-eye with our fellow sentient creatures. Otherwise, as American environmentalist Aldo Leopold points out, those of us who prefer seeing geese to watching television will remain in the minority. And it's about time that I put this thought into action.

The snow is now letting up a bit – a chance to explore beyond the confines of the cosy visitor centre. It hasn't snowed heavily enough to blanket the ground, but it has powdered the grass with white, like a dusting of icing sugar. As I make my way past the dragonfly lakes, with their larvae now sealed beneath a layer of ice, I notice a few birds in the bushes, plucking the last of the berries. Picking them up in my binoculars, I can easily see that they're redwings, more winter visitors from the north. And further down the track is a much larger thrush – a fieldfare – which has also travelled here from Scandinavia, an elegant bird in tastefully coordinated chestnut and grey. In some winters these thrushes are joined by even more elegant birds – waxwings. These birds are dressed all in grey, with a dapper crest, a bright-yellow tail and wing band, and strange red blobs, like sealing wax, on the tips of some of their wing feathers. It's a bird whose name definitely 'does what it says on the tin'.

Waxwings sometimes arrive here in big numbers, usually after a good breeding season in Scandinavia, when their population rises substantially. Then the food runs out. The birds leave the north en masse – a so-called 'irruption' – in their desperate search for more berries. In such years, I've discovered they're far more likely to be found in town than out here on the marshes. They're drawn to the berries of rowan trees, including the commercial varieties now planted widely in parks and gardens. In the past, I've seen flocks of thirty or forty birds in Albert Park, in the centre of Middlesbrough, easily tracked down by their tinkling, bell-like contact calls. Today, many supermarket car parks have been planted with rowan trees and waxwing flocks often spend a week or two in such places as they strip off every berry they can find. These flocks always attract groups of admiring birdwatchers and, drawn by the tangle of tripods and spotting scopes, crowds of curious shoppers.

After a stop to admire the redwings and fieldfares I finally arrive in the hide that overlooks Saltholme Pool. It doesn't feel much warmer inside as the bitter wind howls through the open viewing slits. But there's plenty of activity on the pool to distract from the cold. The winter flocks of wildfowl have arrived to occupy the pools and grazing marshes of Teesmouth in impressive numbers. A huge flock of wigeon is picking its way across the snow-spattered fields. Their slurred, whistling calls, drifting through the snow, are an evocative part of winter marshes up and down the country. These ducks are grazers, and each bird is diligently clipping grass from an exposed patch before moving on to find the next snow-free clump. In this way, the whole flock slowly trundles over the field, further and further from water... and safety.

Then heads go up, and a fraction of second later the flock is airborne, tumbling in an undignified scramble on to the surface of the pool. I scan the sky, suspecting that perhaps it was a peregrine falcon that was responsible for the panic, but I can't see anything, especially through the snow, which is now falling heavily again. If those ducks had spotted a predator (and peregrines are called duck hawks in the United States for a good reason), then they've got better eyesight than me. The wigeon sit there on the water surface, looking indignant – which is something a duck can do extremely well – at the interruption to their lunch. But after about ten or fifteen minutes, they've drifted close to the bank again and a few brave – or hungry – individuals climb out, followed by the rest of the flock. These panic flights to water happen regularly and, sure enough, twenty minutes later, they're off again – plenty of opportunities to capture a few pictures of wigeon flocks in flight through driving snow.

Wigeon nest across the far north of Europe and Asia, though a few birds do nest in Scotland. Around 75 per cent of these ducks winter in Britain, Belgium, the Netherlands and France. That's around a million birds according to a 2016 survey, so it's hardly surprising that favoured places attract such large flocks. These four countries are the heart of the wigeon's winter range and numbers here have remained largely stable over recent years. But the population to the south of this core area has fallen, while numbers of birds wintering closer to their breeding grounds have risen. This is probably a response to climate change, since similar patterns have been noted for many birds, from swans to warblers and thrushes. It's also likely to be the reason for the recent decline of wintering dunlin and bar-tailed godwits on the intertidal mudflats (see Chapter 2). This phenomenon is called 'short-stopping' and makes a lot of sense for the birds. They avoid wasting energy in needless journeys and fly only as far as they have to in order to find an equable winter climate. But a cold snap in the north will soon bring these birds our way.

Scattered across the pool are more than half a dozen other kinds of ducks. In front of the hide, a small group of shovelers has assembled, easily identified by the bright-chestnut sides of the males and, in both males and females, by their outsized, flattened bills. They use these bills, fringed with filtering bristles, to sieve the surface layers of the water for food. They also have a very characteristic – and charming – way of feeding. This small group of two males and two females constantly circle each other, like dodgem cars at a funfair, heads down and bills spluttering through the water. They trace hypnotic patterns over the water's surface as they move slowly across the pool.

Behind them, ever-present mallards are busy upending as they reach down to greater depths to grab whatever they can find that's edible. Across the river, their brethren on the lake in Albert Park are probably feasting on an unhealthy diet of stale white bread, but here on the marshes, they feed in the characteristic way of their kind. As we discovered earlier, Mallards are dabbling ducks. So too are the gadwalls, pintails and teal, which are also feeding on Saltholme Pool today. They all spend a lot of time with their backsides stuck in the air, reaching down for food below the surface. In contrast, the other ducks here are diving ducks, the commonest of which – summer and winter – are tufted ducks.

Male tufted ducks are elegant creatures, formally dressed in black and white, with a trim crest and bright-yellow eye. Females are duller, as is the way in the duck world, although no less sprightly, as they earn their title of 'diving ducks' by plunging underwater. They perform a neat arching leap from the surface, then, pitching in head-first, they disappear in a splash of whirring feet. Ducks are naturally very buoyant, and it takes some effort to remain submerged. Unlike other diving birds, such as auks, ducks don't use their wings underwater – it's all footwork. I've filmed tufted ducks swimming underwater, and they must paddle furiously to avoid bobbing back to the surface like a plastic duck in the bath. It might be comical to watch but it's a serious business for the ducks. Sometimes it's easy to spot a tufted duck's progress, even under murky water, from the roiling turbulence at the surface created by those whirring feet.

There are two rarer diving ducks on the pool today – a few goldeneye and a single female smew – unusual enough to draw a small crowd. Male smew are neat black-and-white ducks. Females, on the other hand, are ruddy-brown, grey and white, but they're just as elegant, and half a dozen spotting scopes are pointed this female's way. Smew belong to a distinctive subgroup of diving ducks – the sawbills, named for the sharp serrations along the margins of their bills that mark them out as fish-eaters, although smew also eat insect larvae when these are abundant. Our two other sawbills, red-breasted merganser and goosander, also winter around Teesmouth, the mergansers more often on the sea just off the coast. But I know I'll almost certainly see goosanders on their nesting territories in spring as I make my way along the middle reaches of the Tees. Both species nest in northern England and Scotland along rivers rich in fish. Although neither is widespread or common, their proficiency at fishing these rivers often brings them into conflict with local human fishermen. Smew, on the other hand, are scarce winter visitors to our shores from northern Scandinavia and Russia.

As I sit in the hide at Saltholme, my joy at these wildfowl spectacles is undiminished by the passing decades. There are ducks everywhere I look, different kinds all going about

124

the business of being a duck in a slightly different way. I've already spent an hour watching wigeon grazing like geese, and witnessed the broadly different feeding styles that mark out dabblers, divers and sawbills. But there are subtler differences. Drifting among the large numbers of tufted ducks, I spot smaller numbers of pochard, another diving duck. Females of these two species look so similar that it might seem that they must be in competition for food, especially since both eat plants and invertebrates. But if food supplies run short, tufted ducks switch their focus almost entirely to invertebrates, while pochards turn to plants, in particular a large branching alga called stonewort. In this way, when the going gets tough, they avoid direct competition.

Scattered among the grazing wigeon, I can see a few lapwings and golden plovers, but the wardens here have told me that there are flocks of many thousands of both of these species hunkered down somewhere out in the rough grasslands surrounding the pools. Golden plovers are close relatives of the grey plovers that we met in Chapter 2 but, unlike this species, they generally avoid the mudflats of Seal Sands and congregate in large flocks on the grassy marshes and pastures.

Some of the lapwings breed locally, in the damp meadows around the estuary or a short distance away in the hills above the Tees Valley, but in winter their numbers are swollen by long-distance migrants from further north and east. Golden plovers, too, breed along the Tees but in their case only on the higher fells and moorland. I'll be able to hear their melancholy whistles and see them in their gold-spangled summer plumage when I explore Teesdale later in my journey. But almost all the golden plovers wintering around Saltholme are visitors from Iceland or Northern Europe.

Both species winter in suitable places right across Britain but in recent years golden plovers' behaviour, like that of wigeon, has changed. Numbers along the east coast have risen, while fewer now winter in the west. This shift might not be immediately apparent if you visit one of the wader hotspots in the west, such as the Severn Estuary, where there are still spectacular flocks of golden plovers and lapwings. But careful analysis of winter counts shows unequivocally that both species are choosing to winter in much larger numbers further east. A similar trend is apparent among some other waders and is yet another manifestation of our changing climate.

It's hard to believe that there are still people, including some in important political positions, who deny the reality of climate change or who don't take it anywhere near seriously enough when there is such abundant evidence all around. Some of this evidence, like the changes in winter distribution of golden plovers and lapwings, is not as obvious as, say, if palm trees suddenly sprang up on the North York Moors, but look at the data and these examples are just as stark a reminder of our very real and growing effects on the planet.

Lapwings and golden plovers on Teesmouth, although occasionally seen on the mudflats, prefer to congregate in tussocky pastures, where they can be surprisingly hard to spot – until they suddenly take to the air. A large flock of either species in flight is a breathtaking sight. Lapwings have broad, rounded wings and fly unhurriedly, like giant butterflies drifting across the sky in slowly changing patterns, conspicuous in their black-and-white plumage. Golden plovers have longer, pointed wings and fly swiftly, often in tightly packed flocks. Their dull backs make them much less obvious than lapwings, but their bellies are bright white. As they twist and turn in unison in a hypnotic aerial dance, the whole flock suddenly shines as thousands of bellies all catch the winter sun.

In the far distance across the pool, I witness a sudden lift-off as thousands of golden plovers and lapwings take to the air in a panic. Surely this time a peregrine must be the culprit, and I rush outside the hide, back into the bitter wind, so I can get a better all-round view. This time I spot the falcon, high overhead – a distinctive dark anchor shape hanging in the sky. As I watch,

it rolls into a steep dive, powered by a few strong wingbeats. Then it tucks in its wings as it stoops almost vertically towards the flock below – at unbelievable speed.

It reaches the panicked flocks in just a second or two and hits a lapwing. The falcon strikes with its feet but doesn't necessarily need to snatch its prey in flight. The speed of the impact kills its victim instantly. A friend told me that on the Tees Estuary he'd once seen a peregrine hit a shelduck with such force that the shelduck's body went one way and its head the other. In the hunt I'm privileged to have just watched, the lapwing remained intact, but tumbled stone-dead out of the sky towards the edge of the pool. As it did, the peregrine twisted gracefully into another dive and effortlessly followed it down. Nature, though, rarely delivers what you expect. The lapwing fell close to a heron, which was intently focused on fishing. The heron was startled from its meditative pose by the unexpected arrival of a dead lapwing and rose up on broad grey wings just as the falcon was about to land next to its prize. The falcon was just as surprised by the sudden appearance of a heron and swiftly rose back into the air. The shock must have banished all thoughts of food because it simply whirled away and headed at speed across the marshes towards the distant Transporter Bridge, leaving the heron bemused – and the lapwing as a treat for the local foxes.

Winter flocks of waders are spectacular as they twist and turn with perfect coordination, perhaps trying to confuse predators like falcons with their mesmeric movement, but the best aerial displays over the marshes at this time of year are performed by roosting flocks of starlings. During the day these birds are dispersed far and wide in their search for food. Large groups almost always hang around the feeders at the centre, making the most of a free lunch, while others feed on a more natural diet in the rough pastures around the reserve. In mid-winter in the North-East it starts to get dark by the middle of the afternoon, and now the feeding flocks begin to merge, gathering in pre-roost assemblies. These groups then begin to converge on the roost site, long lines of birds that coalesce into vast swirling flocks.

If it's not too wet or windy, these flocks put on displays called 'murmurations' in which they perform coordinated aerial manoeuvres. They look like animated smoke drifting across the sky as they create ever-changing patterns. At times, the flock thins and stretches, only to recoil like an elastic band and come together again in a dark churning mass, creating shapes that remind me of living Rorschach ink blots. As with the flocks of lapwings and golden plovers over Saltholme Pool or the swirling masses of knot that I used to see over Seal Sands, it's always been assumed that these displays serve not to impress human observers but to confuse predators. This assumption has recently been tested with a computer game that can generate very life-like behaviour in its digital starlings. Human 'predators' tried to track and 'capture' a single bird, but their performance declined rapidly as the flock size increased, which suggests that it really is harder for a hunter to lock on to a single target in these fluid masses.

A competing theory is that the starlings come together in

such large numbers for warmth on cold winter nights. Most starling roosts are closely observed throughout the winter by an admiring audience, the perfect opportunity for a bit of citizen science to compare these two theories. A group of scientists sent out questionnaires to the flock-watchers, asking them to record the numbers of starlings, their behaviour, the number of potential predators present and the date, weather and temperature. After collating the results, they could find no correlation between flock size and temperature, but the presence of predators made spectacular murmurations far more likely, suggesting that in the real world, as in the computer game, these displays are meant to confuse predators.

The real-life displays are so precisely coordinated that the whole flock seems to turn in perfect synchrony. It's all so slick, it's as if the birds are communicating telepathically but, however it appears, these murmurations are not supernatural events. So how, then, do they do it? Some years ago, when my team produced a film called *Bird Brain*, exploring bird intelligence, we found that starling murmurations are an excellent way to show the raw processing power of bird brains as they coordinate their elaborate choreography. Some of the biggest roosts of all occur in the city of Rome, and scientists there have analysed their swoops and swirls using multiple cameras to generate a three-dimensional model of the position of each bird in the flock. They discovered that each bird locks on to just its nearest seven neighbours, a manageable number, and responds to changes in position of these few birds. Their reaction times are so fast that each manoeuvre spreads rapidly through the flock – so quickly that it seems to the dullard brains of the admiring humans to have been instantaneous.

But training multiple cameras on unpredictable flocks is no easy matter. In the end, the Italian researchers based their observations on much smaller flocks of birds before they joined the main roost, so their behaviour may not have been typical of birds performing their most energetic gyrations at the height of a murmuration. More recently, another group of scientists have suggested that in these large flocks, the birds orientate themselves so that they can see the maximum amount of light through the flock, and as each bird continually adjusts its position, the flock appears to move as one. I like the idea that we still don't quite know how nature puts on such impressive shows. It may not be supernatural, but an air of genuine mystery certainly adds to the spectacle.

In past years at Saltholme, the RSPB has organised 'starling and soup' evenings. Before that gets misinterpreted, let me assure you that starlings are not on the menu. They're the pre-dinner show. Just as the presence of the visitor centre encourages a wide range of people to experience the North Tees Marshes, so 'starlings and soup' gives the folks of Teesside a chance to witness something they might only otherwise see on TV. And although I've spent forty years filming such spectacles for a TV audience, the real thing is far more powerful.

The starling displays often peak in late winter as more birds continue to arrive from further north and east. By the time they finally begin to disperse, winter is almost over. The days have lengthened, and the extra light is triggering thoughts in many creatures of the breeding season to come. And with the returning spring, it's time for me to leave the estuary, its marshes and mudflats, birds and insects, and venture further upstream along the Teesdale Way. The next stage of my journey will take me upstream of the Tees Barrage, and although I will pass through recent urban developments, it will also transport me further back into the history of the Tees. So far, our journey has revealed how the natural world that we see today, even on nature reserves as rich as Saltholme, has been drastically altered or diminished by two centuries of rapid industrialisation and urbanisation, driven by a mindset that sees nature simply as something either to be exploited or simply ignored.

Yet all these changes, which have now brought our planet to the very edge of catastrophic change, have their roots much deeper in time. As we travel along the middle reaches of the Tees, we'll have chance to explore this older history and to give our journey a broader context. Along its middle reaches, the Tees meanders extensively through the landscape, a chance for us to do likewise and follow the twists and turns of history that have shaped the natural and social worlds along the river – and indeed of the whole of Britain.

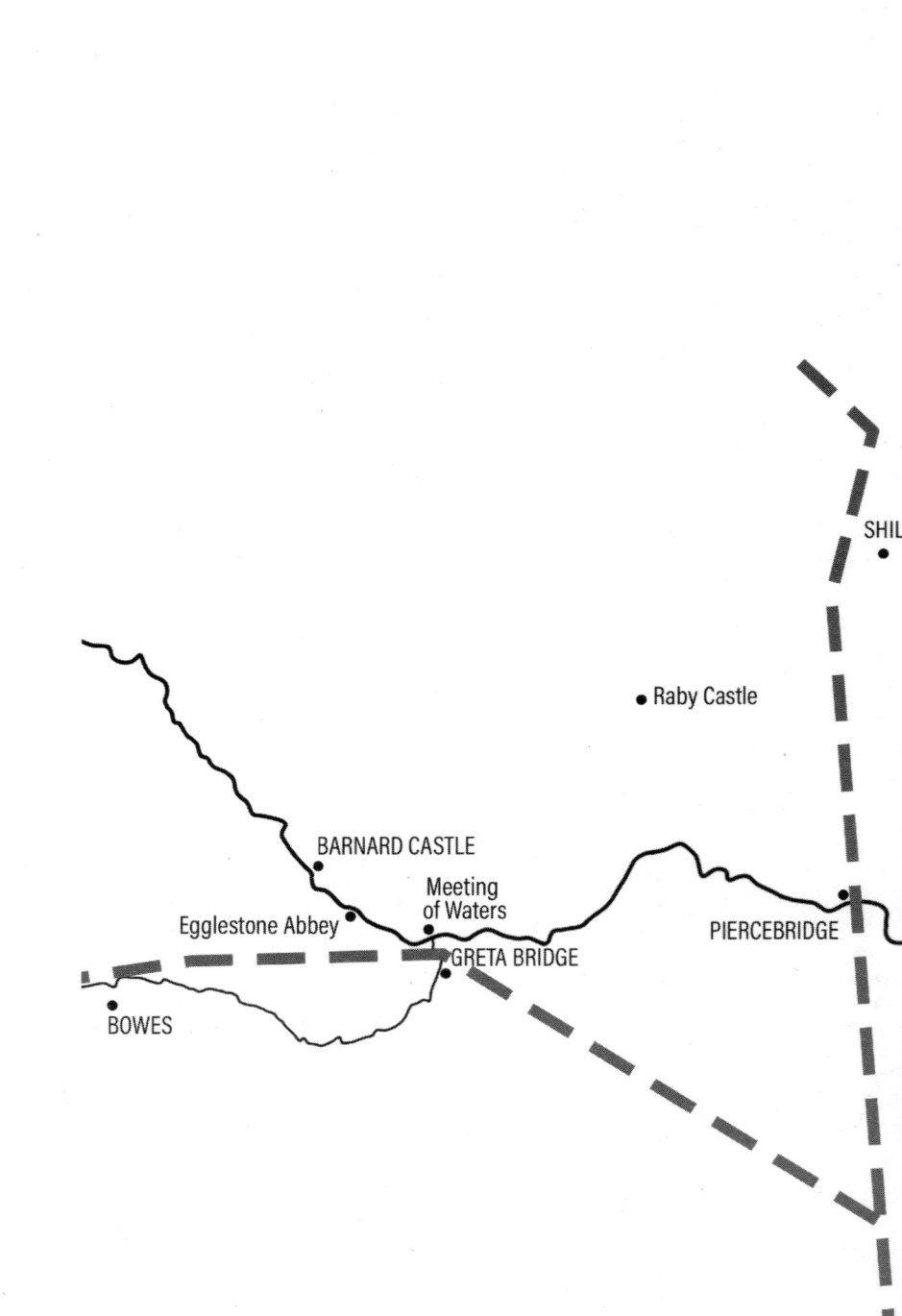

PART II

Middle Reaches

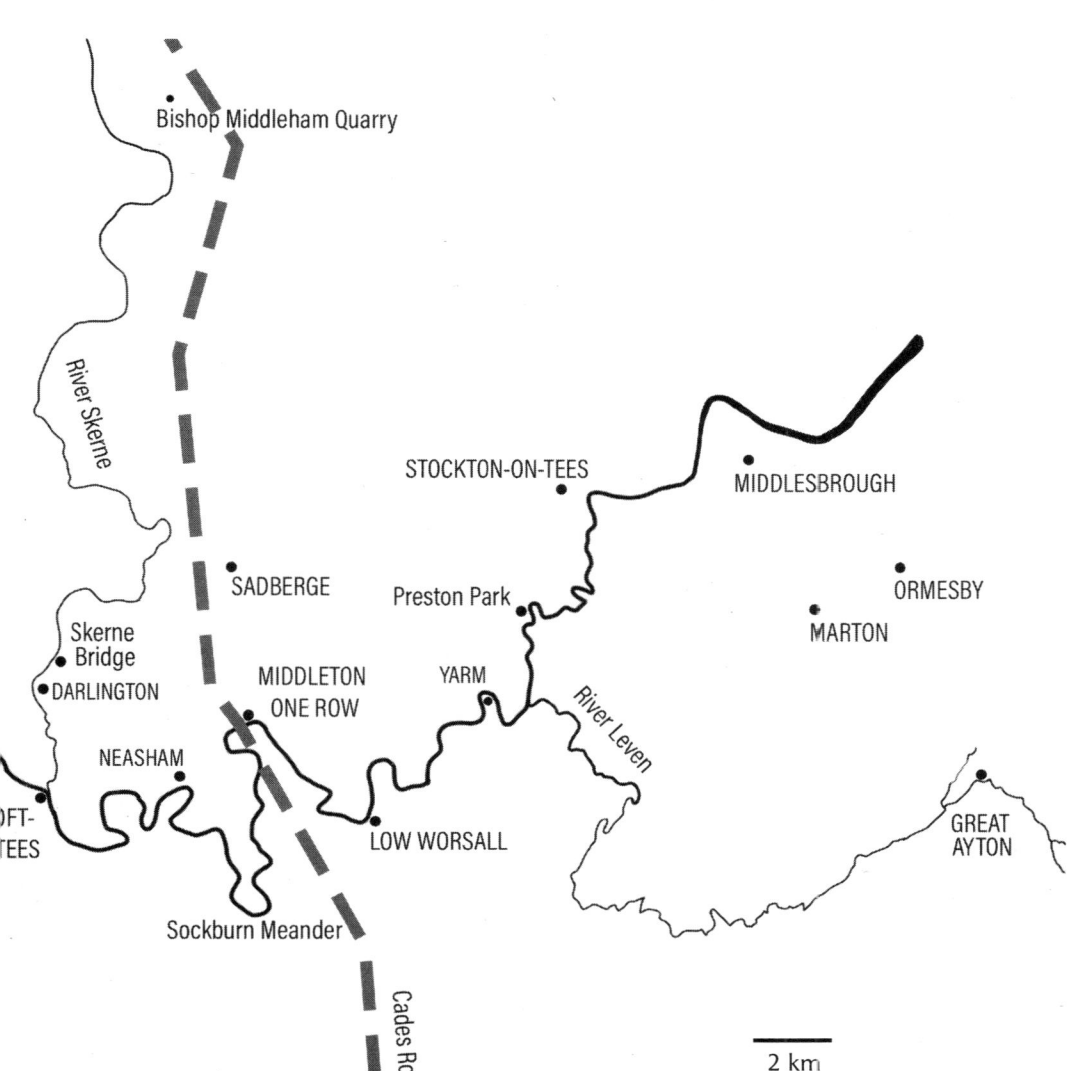

Bishop Middleham Quarry

River Skerne

STOCKTON-ON-TEES

MIDDLESBROUGH

SADBERGE

Preston Park

ORMESBY

MARTON

Skerne
Bridge

DARLINGTON

MIDDLETON
ONE ROW

YARM

River Leven

NEASHAM

OFT-
EES

LOW WORSALL

GREAT
AYTON

Sockburn Meander

Cades Road

2 km

5

The Edge of the World

The Tees Barrage to Yarm – Barriers,
Borders and Boundaries

Ambition leads me not only farther than
any other man has been before me, but as far
as I think it possible for man to go

CAPTAIN JAMES COOK,
HMS *Resolution,* 1774

To boldly go where no man has gone before

CAPTAIN JAMES KIRK,
USS *Enterprise,* 1966

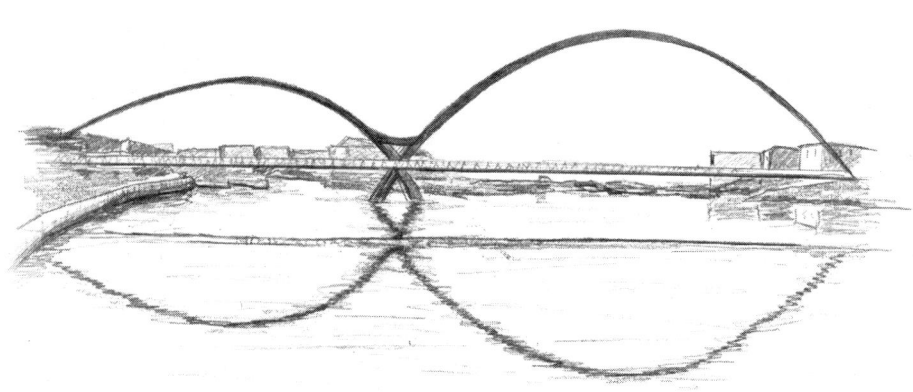

After criss-crossing the wide estuary, our journey now becomes more linear, as the Teesdale Way heads westwards – more or less – following the course of the river – more or less – towards the high country that still lies far beyond the horizon. However, just as tributaries join the Tees along its length and contribute to the river's story, we'll take the odd sidetrack to elaborate on the natural history and the long human history that together make up the story of the Tees Valley. Indeed, this stretch of the river takes us much further back in time than we've travelled so far, beyond the last few centuries that saw the rise of Ironopolis and the transformation of the estuary and lower reaches of the river, to a time of invasions, conquests and ever-shifting borders.

If you hear someone from Middlesbrough saying that they're going 'over the border', they're not referring to a trip to Scotland, or even a venture into neighbouring County Durham. They mean that they're crossing over the railway line that parallels the river's course a little way to the south and east. Over the border is an older Middlesbrough, or at least there was, until successive developments all but obliterated the Victorian heart of the town. This area, between the river and the railways, was where the original town was founded in the early 1800s and later came to be known as St Hilda's, after a now-vanished church that was built to serve the growing community. It lay close to the site of Middlesbrough Priory and to the cell originally dedicated to Saint Hilda by Saint Cuthbert. However, a few buildings, including the old town hall, still stand as reminders of those days. As Middlesbrough grew, the town spread to the south, across the railways, and St Hilda's fell into decline. By the 1960s, 'over the border' had become synonymous with slums.

When I was very young, it really did feel like a world apart – and a far more dangerous one. My parents often took me to Middlesbrough Station to watch trains – big, noisy, huffing and billowing steam trains, of which, quite frankly, I was more scared than impressed. Even then, I was far happier quietly looking for bugs or birds or, even better, collecting treasured conkers in Albert Park.

But the idea of going 'over the border' encapsulates a deeper truth about this whole area. The Tees has been a real border, sometimes a surprising one, for at least the last one thousand years, and probably for a lot longer than that. This shouldn't be surprising. Rivers are broadly linear features in the landscape, conspicuous and often impassable without bridges or ferries, so they make perfect geographical boundaries to mark the edge of a culture's familiar, everyday world. However, anywhere that this boundary is bridged becomes important – the cross-border trade of both goods and ideas often creates wealth for local communities. Many of the towns along our journey, just like a lot of other riverside towns in Britain, began life as bridges or ferries.

So, along this stage of our journey we'll reflect on borders, boundaries and barriers of all kinds – natural and social, geographical and mental – as well as on the importance of crossing some of these boundaries and respecting others. It feels to me that this stretch of the Tees is the perfect place to consider such ideas, locally, nationally and globally, thoughts which occur to me while I'm contemplating the scenery around Maze Park, overlooking the Tees Barrage, on a warm, sunny afternoon.

Maze Park is a recreated patch of nature, built on slag and waste materials from the nearby Head Wrightson site in Thornaby, once a bustling iron- and steelmaking works – one of the first ironworks to be built in the area, although now redeveloped as the Teesdale Business Park. Snaking through thickets of scrub, the grassland is flecked with colour – yellow-wort, black medick, common centaury and bird's-foot trefoil. Early in the year, in May, the abundant hawthorn bushes are covered in such dense blankets of white blossom that it looks as though someone has thrown white tablecloths over them, and the air is heavy with their musky scent. The diversity of plants support a multitude of insects, the most conspicuous of which are the dozen or so kinds of butterflies that have been recorded here. They're joined by several day-flying moths – more burnet moths, like those I found on the damp grasslands near North Gare, and a Mother Shipton moth, less colourful than the burnets, although a closer look reveals its wings to be scrawled with intricate hieroglyphics.

Three steep hills of slag rise above the scrub, and from the top, they allow a broad overview of the area. To the south, on the other side of the wide expanse of railway tracks of the Tees Marshalling Yard, lies the Teesside Retail Park, and to the north-east, the distinctive outline of Newport Bridge. This bridge, a short distance upstream from the Transporter, was opened in 1934 as a vertical-lift bridge.* In other words, the roadway could be lifted on its two supporting pylons to a height of nearly 100 feet above the river – yet more kudos for Middlesbrough. It was the first such bridge in Britain and the largest of its kind in the world. Beyond Newport Bridge is a skyline of cooling towers, belching steam like artificial clouds. To the north, the Tees Barrage and the surrounding park looks a lot more elegant than the other industrial and post-industrial landscapes that encircle Maze Park. It's certainly an impressive development but, as we discovered in Chapter 2, the Tees Barrage is not so much a barrage as a barrier, at least from the perspective of migratory fish.

The barrage is a perfect example of the lack of joined-up thinking that seems to pervade so much of our modern world – in this case where developmental and environmental aspirations were not closely aligned. While a small fortune was spent in developing the barrage and its associated world-class whitewater course, a smaller fortune was being spent on

---

* Since Stockton ceased to be a port for larger vessels, Newport Bridge no longer had to be lifted. So, on 18 November 1990, it was ceremonially raised and lowered one last time.

encouraging salmon to return to the Tees. This included rais-
ing salmon fry and releasing them into the upper reaches of
the river – fish that would have great difficulty in reaching the
sea as smolts and even greater difficulty ascending the river
again as mature adults. It's hardly surprising that migratory
fish populations in the Tees are still a lot lower than those in
other North-Eastern rivers.

However, I'm only singling out the Tees Barrage because it's
sitting right there in front of me, square across the flow of the
river. It's nowhere near the biggest natural disaster that we've
engineered (and are still engineering), but it's symptomatic of
the way our thinking is so compartmentalised – hemmed in by
conceptual boundaries that limit our effectiveness in solving
the myriad problems that we face. Certainly, from my lofty
vantage point in Maze Park, I can see plenty of other places
that encapsulate many decades of both social and environmental injustice along the Tees which stem from a similar lack of
integrated thinking.

The barrage, which opened in 1995, was part of a flurry
of short-term and short-sighted initiatives that originated in

the Thatcher era in response to the collapse of heavy industries along the river and the rise of serious deprivation – a
time when one in five working-age people here were unem-
ployed. Indeed, Margaret Thatcher, the Iron Lady herself,
came to Ironopolis in 1987. There's an iconic picture of her,
clutching her trademark handbag, striding through wasteland
at the former Head Wrightson site, later dubbed the 'walk in
the wilderness'. Locally, it became such a notorious image that
the Middlesbrough Institute of Modern Art (MIMA*) ran an
exhibition in 2017, thirty years after the event, outlining the
impact of a decade of Thatcherism on working-class communities across Britain, from 1977 up to her walk in Thornaby's
wilderness.

I recall that image now, as I survey the vista from the slag
hill – along the river to the Teesdale Business Park, the former
Head Wrightson site, where the Iron Lady strolled through the
wreckage of Teesside's iron and steel industry. The irony of her
epithet was probably lost on her at the time, but this moment
marks an important point in our journey. The history embraced
in the view before me is central to consolidating what we've
discovered so far in the industrial expansion along the lower
Tees, with its social and environmental problems, and in shaping the context for parts of our future journey.

At the start of the 1980s, Margaret Thatcher (prime minister from 1979 to 1990) found an ideological soulmate across
the Atlantic in Ronald Reagan (US president from 1981
to 1989).† It was a relationship and a decade that would

---

* The acronym MIMA bears a close (and deliberate?) resemblance to MOMA
– the world-famous Museum of Modern Art in New York. I smile at the
similarity every time I visit this fine gallery and exhibition space in the centre of
Middlesbrough.
† It wasn't always plain sailing. Like any marriage made in heaven, bits of it
were forged in hell, but nevertheless they pushed the same core agenda out to
the world.

change the world, including this corner of the North-East. Thatcherism and Reaganomics were both forms of the philosophy known as 'neoliberalism' in which maximum economic growth was to be achieved by making the free market as free as possible, through deregulation, the elimination of price controls, the lowering or removal of trade barriers and a shift away from the state and towards privatisation. Together, Thatcher and Reagan promoted neoliberalism on an international scale, with an implicit message that, with high enough economic growth, problems of social inequality and environmental degradation would simply disappear.

Our journey so far shows this to be a false promise. Such 'trickle-down' economics have never worked – the rich always get richer, while the poor get poorer. As of 2015, the world's richest 1 per cent now own 99 per cent of the world's wealth. The Tees has played an important role in nurturing the deep roots of modern neoliberal policies, which lie in the expansion of the British Empire and in the Industrial Revolution of the eighteenth and nineteenth centuries. This revolution was fuelled by coal and later oil, and its infrastructure was built from iron and steel, all commodities that have shaped the last two centuries of life along the lower Tees. However, as we've seen, from the very start, the cycles of boom and bust inherent in this approach have created massive social injustices and severe environmental degradation along the river – problems which the neoliberalism of the 1980s and later has done little to alleviate.

Thatcher's 'walk in the wilderness' was just a couple of years after her battle with Arthur Scargill, leader of the National Union of Miners, underlying which was her fight to limit the power of the unions and their mission to create fairer deals for workers. She appointed a divisive Scots-born industrialist, Ian MacGregor, as head of the National Coal Board to face off with the miners' union. MacGregor had recently overseen a partial 'rescue' of British Steel by reducing its losses, but only at the expense of reducing its workforce by half – the beginning of the end for steelmaking on Teesside. The miners soon faced similar problems as pits across the country were closed. Overnight, Durham mining villages that had helped fuel Britain's Industrial Revolution were turned into ghost towns. Many parts of the country were affected by the policies and philosophies of Thatcherism, but the North-East was hammered by the triple blow of the collapse of steelmaking, the chemical industries and mining. However, the downturn in heavy industry here towards the end of the twentieth century was only the latest in a series of similar events – times when prosperity was closely followed by acutely lean years. In all such downturns, those who suffer most are the working people on whose backs prosperity for an elite few has been built.

Critics of Thatcher saw her walk over derelict land along the Tees as a stroll to admire the final, inevitable result of her policies. However, supporters point to the birth of the urban development corporations that emerged from such visits. These bodies, including the Teesside Development Corporation (TDC), established immediately after her trip to Thornaby, were set up in many deprived areas across the country. Behind these initiatives was a plan to use public money to leverage private finance and to kick-start privatised redevelopment of derelict urban and industrial sites. The TDC clearly had the government's ear, since none of its schemes were ever turned down.

Before the development corporations were wound up in 1998, the TDC oversaw the construction of the Hartlepool Marina, the Teesside Park retail outlets, the Teesdale Business Park on the old Head Wrightson site and, of course, the Tees Barrage. There's no doubt that these developments created opportunities for employment in places where it was desperately needed. For example, I've read that twice as many people now work at the Teesdale Business Park than ever did in the

Head Wrightson works.* Furthermore, the redeveloped area that I can see from my perch on top of one of Maze Park's slag mounds includes a new campus of the University of Durham. So, something to admire has risen from the ashes of big industry along the Tees.

However, the urban development corporations were heavily criticised for riding roughshod over local interests and local people. It was, at its heart, a government-stimulated private-property boom in which investors were the ultimate winners. Indeed, as the TDC was wound up, it was lambasted by Dr Ashok Kumar, the Labour MP for Middlesbrough at the time, who spoke in Parliament to say that the development corporations were 'a glossy and frequently imposing facade hiding a history of often inappropriate and threadbare achievement. The picture is becoming apparent only now, as those bodies come to the end of their lives and the reality of audit and actual achievement are laid bare.'†

Nearly three decades later, little has changed. The latest government initiatives to boost growth in deprived areas are the 'freeports', of which Teesside Freeport is the largest and the first to swing into operation. Freeports are areas exempt from normal tax and customs regulations, designed to boost imports and exports and stimulate the local economy – to the tune of billions, if the press releases can be believed. Environmental considerations have also been relaxed for these developments and those behind the Teesside Freeport have been accused of doing the bare minimum to assess the public health risks of developing such toxic industrial sites. We've already seen that deep dredging at Teesworks, the site of a new deepwater terminal that's part of the freeport network, has been implicated by some scientists in the deaths of huge numbers of crustaceans

---

* The figures I've been able to find don't bear this out. Head Wrightson archives list 6,000 employees, whereas published figures for Teesdale Business Park quote only 4,500 employees.
† Hansard. *Teesside Development Corporation* 304. 14 January 1998.

and other marine creatures. As should now be obvious, in even just the ten miles that we've travelled along the river, there's a long tradition on Teesside of never allowing environmental and public health concerns to stand in the way of making money.

Teesside Freeport is made up of sites scattered along the river, from Redcar and Hartlepool on the coast to Teesside International Airport further upstream, together with several new developments. What's really worrying is that although the freeport itself may only cover around 500 acres, there's a large outer zone surrounding each freeport where a similarly relaxed approach to planning may be applied if the operators of the port can make a compelling commercial case. In the case of the Teesside Freeport, this includes a sizeable chunk of the North York Moors. As of 2023, then-Prime Minister Rishi Sunak's government had failed to clarify the details of this. As the environmental campaigner George Monbiot wrote in an article in *The Guardian*: 'From Teesside to Plymouth, these "special economic zones" are shrouded in secrecy – and exist solely to benefit big business.'[*]

However, right from the start, Teesside Freeport has also been mired in allegations of corruption and illegality over land deals that have greatly benefited a local elite, not unlike the allegations that followed when Thatcher's UDCs were wound up. These latest allegations became so serious that, in May 2023, Michael Gove, then secretary of state for levelling up, housing and communities, ordered an independent review into the affair. However, Gove himself appointed the 'independent' panel. I'm reminded yet again of the story we've already heard, from a century earlier, when, as calls grew in the late 1920s to establish an independent panel to review the sources and effects of pollution along the Tees, industrial magnates pushed to have their own people on the board of inquiry. In the 2023 review,

---

[*] George Monbiot. 'Welcome to the Freeport, Where Turbocapitalism Tramples over British Democracy'. *The Guardian*. 17 August 2022.

the freeport was cleared of cronyism and corruption, although the report was still damning over the secrecy surrounding the financing and value for money of Teesworks, reminiscent of the criticisms levelled against projects promoted by the TDC in the 1990s.

My broad view from Maze Park encompasses examples of all this social and political history – a history in which the ultimate fallacy of past thinking is all too obvious. All our targets for future achievements, as well as the measure of how we're currently faring, are tied to the idea of continued economic growth. It doesn't take a genius to understand that we live on a finite planet – a ball of rock around 8,000 miles in diameter. That's less than the non-stop range of an Airbus A380 jetliner. As the economist Kenneth Boulding told the US Congress in 1973, 'Anyone who believes exponential growth can go on forever in a finite world is either a madman or an economist.' Meanwhile, under this 'business as usual' policy and despite setbacks on the international stage, such as the 2008 financial crisis and the war in Ukraine, the size of the global economy is expected to treble by 2050. It's such policies that have wrenched us from the Holocene and into the Anthropocene, which some academics think would be more accurately termed the 'Capitalocene'.

So, as I gaze down on the physical barrier of the Tees Barrage, I find myself thinking instead of a conceptual barrier – that of our own limited thinking. It's a barrier we must cross if we're to survive in the Anthropocene. One reason for making this journey, for meandering through the natural and social history of the Tees, is to think about how our perspectives must change if we want to make significant improvements in both these realms of life. Since economics lies at the heart of our modern world, our new perspective must be anchored by new economics. History teaches us that such fundamental paradigm shifts, in any area of thinking, are always hard-won, yet growing numbers of economists are now realising that this is exactly what must happen. The visionary Oxford University economist Kate Raworth sees the problem in this way: 'Today we have economies that need to grow, whether or not they make us thrive: what we need are economies that make us thrive, whether or not they grow.'*

Some time ago, I visited a factory complex in Kalundborg, Denmark for a series of films on how nature can inspire new materials and new ways of manufacturing them. It wasn't unlike many of the places we've already passed along the lower Tees, but there was one big difference. The whole place had been set up so that each factory existed in a kind of symbiosis with one or more of the other factories on the site. Waste products from one factory were fed to other factories that could make use of them in their own processes. For example, waste steam from a power station was not vented into the air, as is happening above all the cooling towers around me. Instead, it was used by a nearby pharmaceutical company and an oil refinery. Right around the site, surplus materials or energy from one factory were fed to other factories that could make use of them. It's been operating for five decades, and it works. It's an illustration in microcosm of a form of economy increasingly referred to as circular.

Yet, on this heap of waste slag on Teesside, recolonised by a thriving natural world, I'm surrounded by a system that's part of a global-scale circular economy, one that we are inescapably part of. Among the bright flowers, a carpet of green leaves powered by sunlight is busy building new organic molecules from nothing more than water and carbon dioxide. The only waste product is oxygen, cleaved from water – H_2O – as the plant glues the water's hydrogen atoms to carbon dioxide to make simple sugars. This is the basis of nearly all life on our planet – including our own. We animals then liberate what amounts

* Kate Raworth. *Doughnut Economics: Seven Ways to Think Like a 21st-Century Economist*. Random House. 2022.

to a trapped beam of sunlight by breaking down these sugars – in essence, burning them with oxygen to provide energy (the process of respiration), which releases water and carbon dioxide back into the atmosphere. Plants use up carbon dioxide and create oxygen.* Animals use up oxygen and create carbon dioxide – a circular economy that, with a few ups and downs, has powered the planet for around 2 billion years – and hasn't yet destroyed it.

We can learn a great deal from nature about how to run our own affairs. *Nature Tech*, the documentary series that took me to Kalundborg, was made nearly twenty years ago – and even won an Emmy – but this kind of thinking, although on the increase, is still confined to the fringes of politics and policies, probably because it's new and it's hard, with more long-term than short-term benefits. In other words, it's not so easy to get rich quick. It's reckoned – perhaps optimistically – that possibly 10 per cent of the global economy currently works on such circular principles. Yet, according to the Ellen MacArthur Foundation, moving to a circular model could unlock up to €1.8 trillion of value for Europe's economy.†

Kate Raworth has further developed this way of thinking into a version of the circular economy which she calls 'Doughnut Economics', since the diagram drawn to encapsulate her ideas ends up looking like a doughnut. Raworth's model embeds the more conventional aspects of economics into a doughnut-shaped space, bounded by an outer wall which is defined by the ecological limits of our planet, and an inner wall set by the fundamental rights and needs of society, neither of which form any part of the current neoliberal worldview. Indeed, Thatcher once famously remarked that 'there's no such thing as society'. Other economists are developing a similar approach under the broad

---

* Plants also respire, breaking down organic molecules in the presence of oxygen to release energy and carbon dioxide.
† Ellen MacArthur Foundation. *Growth Within: A Circular Economy Vision for a Competitive Europe*. 2015.

147

heading of 'complexity economics' or 'evolutionary economics', which models economics as an evolutionary system, dynamic and ever-changing, in which economic behaviour is determined by both individuals and society as a whole.* We'll return to this idea at the end of our journey, when we've had chance to amble through the rural stretches of the middle reaches and the wilder landscapes of the upper course of the Tees and after we've had time to explore the long history of the most fundamental aspect of our economy, and the one on which our whole civilisation is based: agriculture.

My musings end abruptly when a small patch of the sandy gravel in front of me detaches itself from the ground and takes to the air. A grayling butterfly. The undersides of its hindwings are coloured and patterned exactly like the bare patches of earth on which it loves to perch, and grasslands growing on slag suit this butterfly very well. I follow this one for a short time, until it lands again, a few feet in front of me, at which point it just vanishes – completely. I took my eye off it, briefly distracted by a flash of intense colour as a common blue butterfly flitted past, and now, no matter how hard I look, I can't see any trace of it. Graylings often tilt their bodies and folded wings away from a potential predator, leaning towards the ground to eliminate any tell-tale shadows, and this one has done a pretty good job of making itself invisible. I leave it in peace and head back down to the barrage, to cross to the north bank and join the path along the river as it heads towards Stockton-on-Tees.

The Teesdale Way leads me past several modern bridges over the river, first the Infinity Bridge, a footbridge so called because, mirrored by its reflection, it assumes the shape of the symbol for infinity. Next comes the Princess of Wales Bridge, a road bridge connecting the university campus and Thornaby

---

* Simon Sharpe. *Five Times Faster: Rethinking the Science, Economics, and Diplomacy of Climate Change.* Cambridge University Press. 2023.

on the south bank with the Northshore housing development. As the river swings south, around the university campus, I pass the Millennium Bridge, another elegant footbridge, before reaching Victoria Bridge, a much older structure and in many ways the heart of Stockton – and of the next part of our story.

This bridge was built between 1882 and 1887 and opened in the fiftieth year of Queen Victoria's reign. Victoria Bridge was the lowest crossing point on the Tees until 1911, when Middlesbrough's iconic Transporter Bridge was built much closer to the river-mouth. However, before Victoria Bridge, there was an even earlier bridge at Stockton, built in 1769. This was to be a toll bridge, at least until the costs of its construction had been met, after which the tolls would be removed. Needless to say, the bridge was raking in cash and the investors were loath to remove the tolls, even after they'd recouped their original costs. So, the good people of Stockton took the matter into their own hands; they threw some of the toll gates into the river and burned the rest on the high street in the town centre. Job done – tolls abolished.

I've already alluded to the key role played by bridges in the history of all our rivers, and Stockton's bridge was no exception. The bridges here, as the lowest crossing point on the river from the mid-eighteenth century until the early twentieth century, prevented large ships from reaching the much older port of Yarm, just four miles upstream – at least as the crow flies (the river itself, and the Teesdale Way, take a much more meandering route). Yarm had been an important port since medieval times, but by the eighteenth century it was being eclipsed by Stockton. I've heard a local expression that succinctly encapsulates the changing relationships of the towns along the lower Tees as the nineteenth century dawned: 'Yarm was, Stockton is, Middlesbrough will be'. In fact, Middlesbrough has rather taken this phrase to heart. The town's motto is 'Erimus', Latin for 'we shall be'. Recalling our journey through the expansive times of Ironopolis, a century and a half ago, it might seem that a more appropriate motto would be 'Fuimus', Latin for 'we have been', which, incidentally, was the motto of the de Brus family, who we'll meet shortly and who came to the North-East after the Norman Conquest. However, I like the sense of optimism for a new future embodied in the current motto. And it's certainly appropriate for the town's football team. I'm still looking forward in anticipation to the glory days for Boro.

Stockton existed long before the de Brus family turned up on Teesside; it dates at least to Anglo-Saxon times. *Stocc Ton* could mean a farmstead of logs (you may recall from Chapter 3 that one suggestion for the derivation of the name of the stock dove was that same word – *stocc*, a log). However, when the word *stocc* appears as the first part of a name, it may also refer to a cell or monastery. Either way, Stockton stood as a settlement, perhaps with a ferry crossing, on the boundary between two Anglo-Saxon kingdoms, Bernicia to the north of the Tees and Deira to the south. In time, these two kingdoms fused into one, encompassing all the lands to the north of the Humber, a kingdom therefore named Northumbria. Unfortunately, this new domain proved unstable. Bickering between the leaders of the two old kingdoms resulted in the occasional murder or assassination, events which predominantly weakened old Deira – at a time when they should have been keeping an eye out for problems from without rather than those from within.

In the late eighth and ninth centuries, the Vikings came. They conquered the part of Northumbria that lay to the south of the Tees and established a strong base at Jorvik (modern York). Most of the lands to the north of the Tees, old Bernicia, remained in Anglo-Saxon hands. Remarkably, this ancient boundary is still apparent today. Towns and villages with Viking place names, ending in 'by', meaning farm or settlement, are very common to the south of the river but are hardly ever found to the north, where Anglo-Saxon 'ton' endings are frequent.* Standing on Victoria Bridge, straddling this ancient boundary, I have Stockton to my left, on the north side of the Tees, and Thornaby to my right, on the south. The boundaries that the Tees has represented through the ages are still there to be seen, just by looking at a map – and it's not just place names. Further upstream, we'll find that the names of many landscape features show a similar division, with a predominance of words of Scandinavian origin to the south of the river, while those of Anglo-Saxon or Celtic origin lie to the north.

For two centuries, Viking culture contributed to the ever-changing British identity, until, in 1066, the Normans arrived. This was, in a way, the last of the Viking invasions of Britain, since the Normans (Northmen) were themselves Vikings who had settled in northern France, where they'd developed their own distinctive culture. In any case, it meant an influx of new people into the North-East, especially barons and other high-ranking Norman nobles whose job it was to control local

---

* 'Thorpe', an outlying settlement, and 'thwaite', a meadow or clearing, are also Old Norse place names which are common to the south of the Tees, while 'worth', an enclosure, is derived from Old English and is common to the north.

populations. The sons of King William I, also known as William the Conqueror,* sent two notable barons to the Tees. King William II (William Rufus) gave Guy de Balliol lands higher up the river, where one of his descendants, Bernard de Balliol, later lent his name to Barnard Castle, so we'll meet him again further along in our travels. A little later, King Henry I granted an estate to Robert de Brus at Skelton, just to the south of where Middlesbrough now stands. Both families would go on to produce kings of Scotland, the more famous being a descendant of Robert's with the same name, Robert the Bruce.

However, long before the de Brus family went to Scotland, Scotland came to them. Less than a century after the Norman invasion, King Stephen gave Northumbria, north of the Tees, to King David of Scotland, so, for a brief period, the Tees became the border between England and Scotland – until King Henry II seized Northumbria back in 1157. In the centuries following the Conquest, it proved hard for Norman kings, based in the South-East of England, to exercise much control over the unruly lands to the north of the Tees, let alone collect taxes. To make matters worse, the old Anglo-Saxon kingdom of Bernicia had two separate seats of power. The lands of the former kingdom were ruled over by the earl of Northumberland, based at Bamburgh Castle on the north-eastern coast, a little to the south of Berwick-on-Tweed. In addition, the bishops of Durham, followers of the Anglo-Saxon Saint Cuthbert, whose earthly remains lie in Durham Cathedral, exercised comparable – if not greater – power.

From the beginning, King William I tried to install loyal Normans into these powerful positions. In 1069, one Norman candidate for Northumbria's earldom, Robert de Comines, made it as far as Durham, where he and his retinue, hundreds

---

* I still smile every time I hear William's title. When, some sixty years ago, I first heard of William the Conqueror, I assumed that, like me, he was a keen collector of prize conkers. The innocence of youth!

152

strong, were summarily slaughtered. An aggrieved conqueror exacted his revenge in a series of attacks on Northumbria that became known as the 'Harrying of the North'. Harried but not conquered, the Northumbrians remained stubbornly independent of Norman control. After much harrying and counter-harrying, a new approach was hatched.

Under William the Conqueror's son, William Rufus, the roles of earl and bishop were combined, creating the prince bishops of Durham, a position then finally filled by Norman clergymen. A substantial amount of power devolved to these new ecclesiastical rulers, as summed up by Anthony Bek, bishop of Durham from 1284 to 1311: 'There are two kings in England, namely the Lord King of England, wearing a crown in sign of his regality and the Lord Bishop of Durham wearing a mitre in place of a crown, in sign of his regality in the diocese of Durham.' The prince bishops had, among other privileges, the power to hold their own parliament, to levy taxes, to administer forests, to mint their own coins and, importantly for the growth of many of the North-East's towns, to create fairs and markets.

However, over time, the stretch of land over which they held sway shrank. Instead of ruling over the whole of the old northern part of the Anglo-Saxon kingdom of Northumbria, running from the Tees up as far as the River Forth, the prince bishop's kingdom stretched only from the Tyne to the Tees, an area that became known as the County Palatine of Durham, practically a separate state that served as a buffer between civilised Norman England and the wild and dangerous lands along the Northumbrian–Scottish border. The bishop's palatine was largely the area we know today as County Durham, its unique heritage proclaimed as you cross the modern county border by signs announcing your arrival in the 'Land of the Prince Bishops'.

Once more, the Tees marked the *de facto* northern border of England and many towns along the river benefitted from this arrangement, Stockton among them. In 1310, Bishop Bek gave

Stockton a market. As Stockton became a centre for trading wool and corn from the fertile valley of the Tees, the town grew. Stockton's heritage as a market town is obvious the minute you stroll down its wide high street, in part actually two high streets separated by a broad marketplace, with the town hall, overlooking all the proceedings, plonked right in the middle.

Before this, the prince bishops had already helped expand the old Anglo-Saxon settlement into a somewhat more substantial place. In the twelfth century, Bishop Hugh Pudsey (served 1153–95) held a manor house at Stockton. Another bishop during the reign of King John, Philip of Poitou (served 1197–1208), is said to have met the king at Stockton. In 1249, Bishop Nicholas Farnham (served 1241–49) retired to the manor at Stockton, when, after just a few years, all the prince-bishoping got too much for him. The manor was rebuilt in 1325 and at some point was fortified to become Stockton Castle, to defend the interests of the prince bishops in the area. In 1470, Laurence Booth (served 1457–76) began a new era for Stockton when he commissioned shipwrights there to build him a ship, the first that we know of in the town. Through the following centuries, shipbuilding expanded, especially when bridges at Stockton began to strangle the river trade to Yarm, further upstream. But, before the birth of modern industrial Stockton, the whole area once again became dangerous border country.

Nearly two centuries after Stockton built its first ship, the Tees once again became the border between England and Scotland. By 1640, King Charles I, with his arrogant ideas of 'Personal Rule', had begun to seriously annoy a lot of people, in particular the Scots – never a good idea. Charles had tried to impose his own version of worship and governance on the Scottish Church, with the threat of excommunication for anyone who argued with the king's supremacy in such matters. In 1637, the introduction of a new Book of Common Prayer was the final straw and riots broke out across Scotland. These soon escalated into all-out war with the Crown.

In the second of two 'Bishops' Wars', the Scottish army defeated the royalists at the Battle of Newburn and by 30 August 1640, they had occupied Newcastle. By 18 September, they had advanced as far as the Tees. Charles was also having a few problems at home, not least with Parliament, which was not enamoured with his 'Personal Rule' either. Unable to raise funds to pursue the war with Scotland, the king was forced into signing the Treaty of Ripon, which established the Tees as the temporary border between England and Scotland. As a final humiliation, Charles was also made to pay living expenses to the Scots while they occupied Durham and Northumberland and even to reimburse them for the costs of the war. He can't have been pleased. These distractions up north emboldened Charles's parliamentary enemies in England, who demanded a larger role in government. On 22 August 1642, Charles declared war on the rebels and England descended into civil war.

By then, Stockton was a critical port. Royalists garrisoned the castle to keep it out of the hands of the Scots and Parliamentarians, but to no avail. The Scots, loyal to the Parliamentarians, took the castle in 1644. In the aftermath of the war, the victorious Oliver Cromwell destroyed Stockton Castle so that no trace of it remains today.

My amble along Stockton High Street, and through a deeper history than we've yet discovered, is a worthwhile sidetrack, since it sets the rest of our journey in a wider context. But the river beckons again. I still have a lengthy trek through urban developments before I can break out into the fertile farmlands of the Tees Valley. But at least those developments ahead of me retain enclaves of a wilder Tees which show yet again how resilient nature can be if it's given half a chance.

In the past, the lower course of an untamed Tees was marked by many long meanders. We've already seen that, in the nineteenth century, cuts made across the bases of two such wide loops straightened the course of the river downstream from Stockton, making a shorter journey for ships to and from the port. Upstream, however, the river still performed its wild gyrations. To the south of Stockton, three long loops of the river originally enclosed agricultural land but, since the barrage was built, preventing the tidal rise and fall of the river, the permanently high water levels have made this area too wet for farming. So, at the start of the new millennium, the Tees Valley Wildlife Trust began restoring more natural habitats of marshes, lakes and wet grasslands.

The first loop encloses a series of reed-fringed ponds that back onto the Bowesfield Industrial Estate, with its roads named after wildflowers – brooklime, yarrow, water avens, sundew – all of which can be seen along the Tees, a few even just down the street on the Bowesfield Marsh nature reserve. Paved paths weave around the pools and through head-high reeds, an easy lunchtime stroll through revitalising nature for those working on the industrial estate. Just as on the pools of the estuary, winter brings flocks of shovelers and teal. Gadwall skulk in the quieter fringes, while ruff and curlew probe exposed mud or wet grasslands. A remarkable 100 species of birds have been recorded on this small patch of rewilded land, surrounded by urban and light-industrial developments.

There are also insects everywhere. The most conspicuous are brown hawker dragonflies, one of our least colourful dragonflies, although their amber-suffused wings make them just as attractive as our brighter species. At the water's edge, I spot a female laying eggs. She's using her ovipositor to saw a slit in a fallen stem just below water level, within which she'll hide her eggs. She's preoccupied with her life's mission, so I can creep close enough to see that, despite first impressions, her brown body is flecked with yellow spots, and she sports two bright-yellow 'go-faster' stripes on each side of her thorax. Over the water, males patrol shoreline territories – animals have their borders too – and even when hidden by the tall reeds, I can hear when a territory-holder has to deal with an intruder. A papery rustle of wings accompanies each aerial skirmish. Finding a vantage point, I watch as two males engage each other in a dog-fight, until both the territory-holder and I realise in the same instant that the intruder is not a rival male. It's a female – and that needs a whole different approach.

If their aerial fights are impressive manoeuvres, their midair coupling is astounding. The male grabs the female with his legs and quickly curls his long abdomen forward. Just like the four-spotted chasers that I watched coupling at Saltholme, a male brown hawker has a pair of claspers at the tip of his abdomen, and he uses these to latch onto the female's eyes. Now the pair fly as one, in tandem. They head into a nearby willow tree, where they perform the same gymnastics as the four-spotted chasers, the female bending her abdomen up to connect with the male's secondary sex organs at the base of his abdomen. Time to give them some privacy.

The footpath eventually takes me to the bank of the river, where I spot several male banded demoiselles, large damsel-flies that prefer slow-moving water. The Tees at this point in its journey to the sea is a perfect home for them. There are several males, with iridescent blue and green bodies, perched on over-hanging vegetation, wings clasped together like sails over their backs. Each wing is crossed by a broad band of deep, inky blue that seems to glow in the bright sunshine. They're stunningly beautiful insects. These males are also territorial, although each defends a much smaller patch than the brown hawkers on the lakes behind me. And they're less inclined to fight – unless they have to.

Their boldly marked wings serve as banners to display to rivals. The extent of the blue bars varies between individuals – fitter, tougher males have broader ones. If a male intrudes on another's patch, all the resident needs to do is display his status by flicking his wings. If the intruder has smaller wing bars, he won't risk a fight with an inevitable outcome and he retires gracefully. However, if two males have similar-sized wing bars, the contest escalates. They fly out over the water using exaggerated slow wing beats, gyrating around each other. If that doesn't settle the dispute amicably, the two males engage in a brief fight. Usually in nature, possession is nine-tenths of the law and the resident wins, unless his opponent is a lot tougher than him – in which case, it's off to find a quieter spot further along the river.

Following the winding course of the river brings me to the Horseshoe Bend, a loop enclosing The Holmes, yet another name of Viking origin, on the southern side of the river. In Old Norse, holm referred to an island in a river, which this peninsula almost is, bounded on three sides by the river. The third bend encloses another large lake, while on the opposite bank lies Bassleton Woods, a small remnant of the ancient woodland that once fringed much of the lower Tees. All these sites on each side of the river are linked by all-weather footpaths as part of the Tees Heritage Park.

The Tees Heritage Park was launched in 2007 and encompasses the river valley between Stockton and Darlington. Its mission is to preserve the social and industrial heritage along this stretch of the river and to enhance its natural heritage. The network of footpaths allows easy access to many features of historic value as well as to natural habitats, from lakes and marshes to grasslands and ancient woodlands. At first glance, this section of the river seems entirely urban or industrialised, but look closer and you will discover many enclaves where wild creatures and plants still thrive and where we humans can refresh our souls – and, even amid all this development, there are still places along the river where you can stand entirely encircled by nature. Here is yet another example of how little it takes for the natural world to reclaim lost ground.

Although there are Tees Heritage Park footpaths along the river, the official Teesdale Way now takes a short overland detour. It runs briefly along the busy Queen Elizabeth Way, the main road linking Stockton to the recent large housing development at Ingleby Barwick. Thankfully, it soon veers off to cross Preston Farm at the top of a slope that presents a wide view of the Tees. Below, the river is crossed by a collection of pipes, a reminder of the once-extensive industry that ran right along this stretch of the river. On the top of the slope, a newly ploughed field has drawn flocks of rooks and jackdaws, eager to see what morsels have been turned up.

The path soon enters a narrow strip of woodland and once more descends to the riverbank, again blotting out the modern world. In the autumn, this section of the river can be magical. In calm, misty weather, there's not a ripple on the surface of the water, making it the perfect mirror for the soft-golden-yellow trees, creating a flawless symmetry. In spring, the forest floor is a carpet of white ramsons, flowing right down to the banks of the river, and the trees are noisy with the lively calls of great and blue tits, and the sibilant whistles of nuthatches.

A short walk through this respite from the hustle and bustle brings me into Preston Park, now a museum and the centre of the Tees Heritage Park. Like Stockton, Preston-on-Tees itself is an ancient settlement. Since the Tees in Norman times was effectively the northern boundary of rule from London, those places to the north of the river were not surveyed in William the Conqueror's Domesday Book, a general survey of the country completed in 1086. However, Preston is mentioned in the Boldon Buke, County Durham's own version of the Domesday Book, completed nearly a century later, in 1183.

This separate survey was commissioned by Hugh Pudsey, bishop of Durham. Like the earlier Domesday Book, it was essentially a survey of who owed what taxes, although in this case they were payable to the bishops of Durham rather than to the Crown, as was the case to the south of the river.* Through the ages, Preston passed between different families until the site of Preston Park became the property of David Fowler in 1812. In 1825, the same year that the Stockton and Darlington Railway opened, Fowler built Preston Hall on the site. The railway ran through Preston Park, where today there's a footpath along the route, past a few unprepossessing mounds that were once part of the original track bed. More impressive, the museum houses the Yarm Helmet, a direct link to the Viking history of the Tees. It's the first ever complete (or nearly complete) Viking helmet to have been found in Britain and has been dated to the tenth century. It was found in the 1950s by workers laying new sewer pipes in Yarm and owes its remarkable preservation to the waterlogged soils close to the Tees where it lay buried for a millennium. I'm captivated by this rusty old helmet since I can't help but wonder who wore it and what stories they could tell of the Tees Valley a thousand years ago. Objects like the helmet bring these nameless people close enough to touch.

After Preston Park, the Teesdale Way heads away from the river and joins the busy main road from Stockton to Yarm for about half a mile before it cuts back towards the river, skirting the Eaglescliffe golf course. A short walk along the left bank brings me to the confluence of the Tees and the Leven on the opposite shore. The Leven rises on the north slopes of the North York Moors before making its tortuous way through Great Ayton and Stokesley to its union with the Tees. Like the Tees, it was once full of fish, but a weir built across the lower river during the Industrial Revolution was as effective as the Tees Barrage in blocking the routes of fish that migrate from river to ocean and back again.

In 2007, a fish pass was built around the weir at Leven

---

* There are a number of 'enclaves' of the County Palatine of Durham south of the river in what is now Yorkshire, for example, around Crayke, near York, which therefore also paid taxes to the prince bishops.

Bridge, and in 2011, salmon finally made it back to their ancient spawning grounds, the first to spawn in the Leven since the eighteenth century. Sea trout also now move upriver again, although I feel for these fish that must first make it past the Tees Barrage and then ascend the fish pass on the Leven. Higher up the Leven, they mingle with fish more characteristic of the many smaller, stony streams on the other side of the North York Moors watershed that drain into the Esk, which follows its own route to the North Sea at Whitby.

The River Leven at Great Ayton was only a bike ride from my childhood home, so it was a frequently visited spot (all the more enticing since Suggitt's Ice Creams sits right next to the river). In places the river is shallow enough to wade, to turn over stones on the riverbed and discover what lurks beneath. I still find it hard to resist this pastime (and ice cream). Sometimes a stone loach darts away to safety. Occasionally, I find a yellow mass of eggs clinging to the underside of a stone. These are the eggs of a bullhead, a fish sometimes called the miller's thumb, owing to its stumpy, flattened shape, like the thumb of a miller flattened by years of rubbing corn meal between finger and thumb to gauge its quality. The male bullhead excavates a small cave beneath a stone and entices a female to visit. She lays her eggs on the roof of his cave and the male guards them until they hatch. So, I'm always careful to place the stone back exactly as I found it.

But the most exciting find today is a fish that could easily be mistaken for an eel, although it couldn't be more different – a brook lamprey. Lampreys are jawless fish which resemble some of the very earliest vertebrates from about 500 million years ago. Like lampreys, these creatures had rows of gills along the front part of their bodies. Some of these ancient fish later evolved jaws by repurposing the structures that supported their first gill arch. That's how all vertebrate jaws arose, from the gaping maw of a great white shark to my own, busy crunching up the remains of a Suggitt's ice cream cone while I admire the lamprey I've just caught. It's all gill and no jaws – but it does have a strange sucker-like mouth at its front end.

Lampreys also have a peculiar life cycle. The young stage of a lamprey's life, when it is known as an 'ammocoete' larva, lasts several years, during which time it lives buried in mud and feeds by filtering tiny particles through its gills. The larva grows to the size of an adult and could easily be mistaken for one, except for its water-filtering lifestyle. The larva then metamorphoses into a sexually mature adult which breeds in a similar way to salmon, in gravel spawning beds. It uses its sucker-like mouth to attach to a stone, which it drags away to reveal a bed of fine gravel into which it lays its eggs. Males and females can sometimes be seen entwined around each other in a sensual embrace as they release eggs and sperm.

During their adult lives brook lampreys don't feed, surviving on food reserves laid down as ammocoetes. However, two other species of lamprey live in Britain's rivers and they add a macabre twist to the story. When they become adults, river and sea lampreys develop rings of sharp teeth around their suckers and switch from filter-feeding to a diet of flesh and blood. Unlike brook lampreys, which spend their whole lives in the river where they hatched, river and sea lampreys migrate downstream and out to sea, where they use their suckers to attach to marine fish or even whales. They rasp away at flesh using the circular saws inside their suckers, often leaving distinctive scars when they detach. Sea lampreys reach impressively large sizes, sometimes over three feet in length and weighing more than five pounds. When they were a lot more abundant, lampreys of all kinds were valued as food. Indeed, it's said that

King Henry I, fourth son of William the Conqueror, died from eating a surfeit of them. Today, it's hard to find anything even remotely resembling a surfeit of these intriguing fish. River and sea lampreys face all the same problems as other migratory fish as they negotiate barriers to their migration such as weirs and dams, or severe pollution in the lower reaches of rivers. Sea lampreys are often recorded on the downstream side of the Tees Barrage, although, lacking a salmon's athleticism, they find it all but impossible to cross this barrier.

Leaving no stone unturned, I soon see a crayfish make its escape with a powerful flick of its tail. The native white-clawed crayfish still lives in the Leven but, like crayfish across the country, is under threat from introduced species from North America. Many different species of crayfish abound in North America, both in and out of the water. I've walked through swampy bottomlands along the Mississippi, where the forest floor is covered in the characteristic mud turrets built by some kinds of crayfish. The turrets mark the entrances to burrows in the damp soil that may reach down fifteen feet or more. These little creatures exist in such abundance that they're important ecosystem engineers, improving soil structure and aeration for plants as well as being an invaluable source of food for both local wildlife and local people. They are, after all, a key ingredient in Louisiana's signature dish, crawfish gumbo.

Some of these abundant and larger crayfish were imported into Europe as long ago as 1859 to bolster supplies of native species but they brought with them crayfish plague, a disease to which North American species are immune, but which kills the white-clawed crayfish in just a few weeks. The disease is easily introduced to new areas on waders and boots or fishing gear that's been used on an infected waterway – in this way it has spread rapidly throughout the UK. The disease is also spread when North American crayfish invade new river systems, a bleak prognosis for our native crayfish. Some predict that the white-clawed crayfish could be extinct over most of its British range within twenty years. But we can all do our bit. When I get back to the car, I peel off my waders and store them in a large bag. At home, I'll rinse them down and let them dry before I use them again – better safe than sorry.

River creatures and ice cream are not my only reasons for visiting Great Ayton. Several branches of my family tree have a long history in this village, going back at least to the early 1700s. The nineteenth-century Great Ayton Cemetery, with views across to the iconic outline of Roseberry Topping marking the northern edge of the North York Moors, is the final resting place of several generations of Nicholls (or Nichols). However, there's also a much older church here – All Saints Church – which was built in the twelfth century. Before that, in 1086, the Domesday Book records an Anglo-Saxon church on this site which seems to have survived the Harrying of the North, a time when many Anglo-Saxon churches in the area were destroyed. In this churchyard there are graves of several Bulmers, another ancestral family, although these two lines, both long-time residents of Great Ayton, didn't meet until my father and mother married. All Saints is a beautiful little church, built from local sandstone, and a peaceful place to wander through my family history. But there's another grave here, closely flanked by the Bulmers that I want to visit. It belongs to Grace Cook, the mother of Captain James Cook – another Middlesbrough lad (or he would have been if Middlesbrough had existed when he was born in 1728).* Nevertheless, Cook would play a significant role in the later evolution of Middlesbrough and the lower Tees and in the birth of the economic philosophies that we mused over at Maze Park at the start of this leg of our journey.

I feel a close connection to this explorer, navigator and map-maker extraordinaire because, for my whole life, I've kept crossing his path, both in the North-East of England and elsewhere

---

* Cook's father is buried in Marske, a small village just down the coast from Redcar.

around the planet. While researching a film project, I recently visited the UK Hydrographic Office in Taunton, Somerset, and although Cook wasn't the subject of my research, I nevertheless couldn't resist asking my host if I could see some of Cook's original charts, which are kept there. He duly obliged. I knew Cook's charts were far in advance of their time and remarkable for their accuracy and detail, but seeing the real thing underlined what an extraordinary achievement they are, created from the deck of a rolling ship while travelling along unknown coasts, often occupied by native peoples keen to discourage these strange visitors with all possible vigour. The maps are beautiful works of art and yet so accurate that some remained in use well into the twentieth century. David Attenborough recalls using Cook's charts of the Great Barrier Reef, off the Queensland coast, on one of his own early visits to the area. To judge from Cook's own words, with which we began this chapter, he was a man who lived to cross boundaries.

Cook's story begins very close to my own. He was born less than a mile from where I first entered the world. Before his birth, Cook's family lived and worked in Ormesby, then a small village, now absorbed into the conurbation of Middlesbrough, although still with its own character. I went to infant school there and my parents are buried in the churchyard of St Cuthbert's in Ormesby. The Cooks then moved to Marton, another small village just over a mile to the west. Here, in a small cottage, the navigator James Cook was born. When, as a young child, I began to read about Cook, I discovered that I spent the first ten years of my life living just half a mile from where Cook grew up for his first eight years. As adventurous kids, we doubtless explored the same ground. Eventually, the Cook family moved to Great Ayton, to become part of the small community that included several of my own ancestors.

Cook's life story is well known and documented in numerous books, the best of which I've listed in the bibliography. He learned his seamanship plying the stormy North Sea from the port of Whitby on the coast of North Yorkshire. He then joined the Royal Navy, where his skills as a seaman and navigator saw him rise quickly to command three trips to the furthest reaches of the globe. During my own decades of travel, my path has crossed his on numerous occasions, from the West Coast of North America and the remote Hawaiian archipelago to Australia, where Cook is even more celebrated than he is in the North-East of England.

While filming in Queensland, I visited the curiously named town of Seventeen Seventy, so called to celebrate the year that Cook landed there, his second port of call in Australia after landing in Botany Bay, near Sydney, in New South Wales. There's a cairn in nearby Bustard Bay that commemorates the event, standing on the site where one of Cook's men carved his name on a tree. On a later trip, filming in the magnificent rain-forests of Daintree in Far North Queensland, I found myself at Cape Tribulation, so named by Cook since this is where some of the worst problems on his first voyage began.

Offshore, he'd had a rather-too-close encounter with the Great Barrier Reef, which stretches for well over 1,200 miles along this coast, and so is quite hard to miss. It's only down to Cook's seamanship that they didn't run aground more often in these uncharted waters. However, on this occasion, his ship, the HMS *Endeavour*, was badly damaged. The reef on which he foundered is now called Endeavour Reef, and his brush with the hard, sharp coral just below the surface holed the *Endeavour*. Somehow, Cook was able to negotiate the maze of narrow, coral-lined channels to limp into a broad river, now called the Endeavour River, where he could carry out repairs.

This man from an obscure village in the valley of the Tees mapped about a quarter of the globe and the naturalists he took on all his voyages greatly extended our understanding of the natural world. His biographer Sir Walter Besant said of him, 'No other sailor has ever so greatly enlarged the borders of the earth.' For his time, he was also an advanced thinker. He remained humble and polite, though firm in his convictions and his command. While Cook tried to foster peaceful relationships with the indigenous peoples he met,* the voyages he made were essentially imperialist, and would later result in lands being stolen from their original inhabitants, from the West Coast of America to Australia and New Zealand. Cook regretted this. In the captain's log for his second voyage, he recorded how sorry he was to see natives, who had been living simple yet happy lives, being deprived of their land and their ways of life by cultures with a very different worldview.

Colonialism and the subsequent rise of empire laid the foundations for the rapid economic growth of Britain from which the Industrial Revolution would eventually emerge. So, Cook, in mapping so much of the world and cataloguing the resources he discovered, played a big part in this process and in the later development of the lower Tees where he was born. Exploration and colonialism in the seventeenth and eighteenth centuries saw the rise of 'mercantile capitalism', based on stripping assets, both natural and manufactured, from colonised countries. Out of this would emerge 'industrial capitalism', which triggered the growth of cities and an urban working class. Beginning in Britain in the latter part of the eighteenth century, within half a century industrial capitalism had spread across Europe. It then flourished throughout the nineteenth century in places such as the lower Tees Valley and further afield, accelerating greatly the practice of fossil-fuel burning that has now brought the world to the brink of catastrophic climate breakdown. In a post-industrial age, 'consumer capitalism' – which has generated a relentless cycle of demand for, and over-consumption of, finite resources – is now wreaking its own brand of havoc.

Captain James Cook RN played no small part in this story. Nevertheless, in a time when we're reassessing many of the heroes of our past for their imperialism and their roles in such

---

* Cook would, however, meet his death in 1779, during his third voyage of discovery, at the hands of Hawaiians, after relations between Europeans and islanders broke down.

activities as slavery, although he is not without his faults, Cook still stands as a worthy icon of science and exploration – of venturing across borders, into the unknown, whether physically or in spirit. In fact, Cook's legacy has gone much further than even he could imagine travelling, reflected in the reason that I picked the two disparate quotes at the start of this chapter, one from James Cook, the other from the suspiciously similar-sounding James Kirk, a fictional captain in the *Star Trek* series. Both quotes express the same idea, and I wouldn't be at all surprised if the writers of *Star Trek* were familiar with Cook's writing. Yet, more than that, Cook's legendary voyages are commemorated in a real spacecraft. Space Shuttle *Endeavour*, the last shuttle to fly, was named for Cook's first ship. Incidentally, the prototype shuttle was named *Enterprise*, after Kirk's starship.

It's time to conclude our foray along the Leven and into the life of one of Middlesbrough's most famous sons (or at least adopted by Middlesbrough when it sprang into existence nearly a century after his birth). It's a short journey back to the Tees from Great Ayton, a journey also undertaken by a few of my ancestors. Several Nicholls moved to Middlesbrough to work in the burgeoning industries there, whereas my third great-grandfather, William Bulmer, moved to Stockton. He was a stonemason, like his father, and set up a stonemasonry business on Stockton High Street. In his late sixties, he changed his property at 163 High Street into a public house, the Cleveland Hotel. After he died, the pub changed its name to the North Eastern Public House, but eventually the old building was torn down. Having been rebuilt in the late nineteenth century, the 'new' North Eastern pub still stands on Stockton High Street.

From the confluence of the Rivers Leven and Tees, it's a short hop to our next port of call. Enclosed by another large meander of the Tees, the town of Yarm is a place with a very different story to tell. From Yarm upstream to Croft-on-Tees, we'll encounter some monster fish stories along with some very real monsters – and a world-famous grinning cat.

6

River Monsters

*Yarm to Croft-on-Tees – Salmon, Sturgeon and
the Sockburn Worm*

> Beware the Jabberwock, my son!
> The jaws that bite, the claws that catch!
> Beware the Jubjub bird, and shun
> The frumious Bandersnatch!

LEWIS CARROLL, 'Jabberwocky', 1871

Yarm, or at least old Yarm, nestles in yet another long northward loop of the Tees, which encloses the town on three sides. The Teesdale Way follows the outside curve of this meander through open woodlands with views across the river until, at the apex of the bend, it emerges, providentially, in the grounds of the Blue Bell Inn. Well – don't mind if I do... Next to the pub is the old stone-built Yarm Bridge – centuries ago, the first bridge that I'd have encountered on my journey upriver. Even though Yarm is just a few miles from Stockton and the urban and light-industrial sprawl that has spread through Eaglescliffe right up to Yarm, crossing the ancient stone bridge into the old town is like stepping back in time. So, after a refreshing pint, I stride with renewed vigour into Yarm's long history.

Yarm could be a town set in some remote corner of rural Britain. Across Yarm Bridge, I'm soon wandering down the broad high street, comparable to that of Stockton, which marks Yarm as another town that prospered around an old market. Yarm's market, however, existed long before that of Stockton and the old town has retained an air of times long past, notwithstanding the ranks of trendy coffee shops and bars, souvenir shops and fashion shops that now occupy the old buildings. But a settlement existed here even before the market came to town – and the clue is in the name.

'Yarm' is derived from the Old English *gear* (pronounced 'yair'), meaning a fish weir. The plural form is *gearum*, hence Yarm was a place of fish weirs which were built to trap fish moving with the tides that flowed through Yarm before the Tees Barrage was built. From earliest times right up until the middle of the nineteenth century, fish abounded in the Tees. As we've seen, it was once one of the best salmon rivers in the country and along with these valuable fish, the waters of the Tees teemed with bream, chub, eels, grayling, trout and pike, along with smaller species, such as stone loaches, bullheads, three-spined sticklebacks and gudgeon, of less interest to subsistence

or market fishermen, although of great interest to me. Over the years, I've kept all these species in aquaria to better make their acquaintance.

There are still some impressively large fish in the river around Yarm – bream and pike, for example, that reach twenty to twenty-five pounds – but I'm certain that Anglo-Saxon fishermen would have pulled some real monsters out of the river here. Some years ago, in researching my book *Paradise Found*,* on the history of North America's natural world, I had to scour old documents, written by the earliest Europeans to explore and settle the continent, for descriptions of the animals and plants that they found. The book was written to highlight the dramatic fall in the abundance of nature, as opposed to the more familiar concept of the loss of biodiversity, and my research revealed that nature was far more abundant in the past than most of us would ever believe.

Although North America had been occupied by people for at least 15,000 years, and perhaps a lot longer, it was only in the

---

* Steve Nicholls. *Paradise Found: Nature in America at the Time of Discovery.* University of Chicago Press. 2009.

late fifteenth century that Europeans began to explore this new world. They left descriptions of an extraordinary natural world which allowed me to piece together a picture of North America before it was diminished by rampant over-exploitation – as a nascent free-market economy, first in Europe and then later in North America itself, began to erode nature's abundance. The decline in biodiversity is an easy – if harrowing – concept to grasp. You know with certainty that you'll never see a dodo, a great auk or a Labrador duck. You know you'd search in vain for Carolina parakeets or Cuban macaws, Steller's sea cows or thylacines. It's a binary concept – it's either alive and kicking or it's gone forever.*

On the other hand, 'bioabundance' is a much harder idea to grasp intuitively. It's quantitative rather than qualitative and it's subject to a phenomenon called 'shifting baseline syndrome'. When I wrote *Paradise Found* in 2009, this wasn't a famil-iar concept, although it's been more widely discussed in recent years. Imagine the natural world that surrounded my great-great-grandparents in their youth. Over their lives, as nature inevitably retreated, they witnessed the decline in abundance, and at the end of their lives, recalling their idyllic childhood, they would have had a very real sense of how much had been lost. Their children too, my great-grandparents, were surround-ed by nature, though less than their parents. They, too, watched its decline and at the end of their lives, recalling their own child-hoods, had a clear sense of what had been lost, but the starting point for their personal experience was different from that of their parents; nature was already less abundant. The baseline from which they judged what had been lost had shifted so that they inevitably underestimated the true scale of the loss.

* Like so many things in nature, it's never quite that simple. Sometimes species declared extinct turn up again – so-called Lazarus species. For example, the Bermuda petrel, or cahow, is a seabird that nests on Bermuda, where it was presumed extinct for over 300 years until intrepid ornithologists discovered a small colony on a remote islet in 1951.

Now imagine that same process repeated by my grandparents and my parents and then by me. I certainly recall a childhood where insects and birds were far more abundant than they are now, a subjective memory that's backed by some alarming figures, since over the last few generations we've been recording natural abundance more scientifically. Since I began chasing bugs and birds in the late 1960s, 80 per cent of our butterflies have suffered serious declines.[*] Over the same period, farmland birds have declined in abundance by 60 per cent.[†] The *State of Nature* report for 2023 concludes that today, one in every six species of animals and plants in the UK is in danger of extinction. But even more importantly, the baseline for both my own judgements and these national statistics had, by the late 1960s, already shifted a very long way. Thanks to the work I did in researching *Paradise Found*, I know that neither I nor anyone else alive today has any idea of what pristine nature really looks like. If we ever invent a time machine, when we step outside into a past world, we will be dumbstruck by the sheer abundance of the life around us.

The records for North America began in 1492, with Columbus, although most are more recent and more detailed. The loss of natural abundance happened much earlier in most of Britain, so the records here are less complete. Even so, some idea of the abundance of fish in English waters in Anglo-Saxon times can be gleaned from an old document which reveals that, in the 970s, 20 fishermen gave 60,000 eels to the monks at Ramsey Abbey in Cambridgeshire as a form of fishy tax. The fens continued to produce enormous quantities of eels in the following centuries. The Domesday Book records 24,000 eels paid as taxes in Stuntney and 27,150 in Doddington, also in Cambridgeshire. Although such 'renders'

[*] Butterfly Conservation. *The State of the UK's Butterflies.* 2022.
[†] See the British Trust for Ornithology's Farmland Bird Index. https://www.bto.org/our-science/publications/developing-bird-indicators#farmland

of eels were smaller in other parts of the country, it's clear that even at the start of the Norman period, fish were often to be found in great abundance. Today, it would be hard to catch anything like this number – even if His Majesty's Revenue and Customs could be persuaded to accept eels as payment for taxes. I don't know of comparable data for the Tees this early in its history, but the records from elsewhere hint at the scale of the harvest available from the Anglo-Saxon fish weirs at Yarm. Yet today, eels are endangered not just on the Tees, but throughout much of their range.

Smaller fish also existed in numbers we'd have trouble imagining today. In 1808, Captain T. Williamson, in his book *The Complete Angler's Vade-Mecum*, wrote that sticklebacks were 'sold by the bushel, as manure, both in Lincolnshire and in Cambridgeshire'. In 1815, Thomas Salter, in *The Angler's Guide*, suggested even greater quantities. He reported that 'pricklebacks' sometimes choked the River Welland, 'at which time they are collected in nets, sieves, baskets &c., to the amount of cart loads and spread on the land as manure'. As a keen observer and collector of sticklebacks in my early childhood, I can hardly imagine such extraordinary sights – certainly more than I ever found in the lake at Albert Park.

Fish-watching in the past would have also held another thrill – the rivers of Britain hid some real monsters. Many creatures, like fish or reptiles, carry on growing throughout their lives, and grew to much larger sizes in the past, since before heavy exploitation they lived for a lot longer. This is also true of molluscs, which explains why the first settlers to colonise the Chesapeake Bay, on America's East Coast, described finding oysters over a foot long, so large that a single one was enough to feed four men. Reports from elsewhere suggest that many creatures were not only far more abundant in the past, but similarly grew to much larger sizes – from cod and salmon to sharks and sea turtles. So, I'm sure that the fishermen of Yarm would have encountered plenty of such monsters in the past.

More and bigger fish meant that even in the nineteenth century the Tees at Yarm provided a bounteous harvest.

The Blue Bell Inn where I quenched my thirst on my way into Yarm was serving ale to local fishermen back in the middle of the nineteenth century. Indeed, at the time, its landlord George Goldie was as well known as a commercial salmon fisherman as he was as a publican – although he would be the last in a long line of people to make a living by hauling salmon from this stretch of the Tees. In a good year, he ran six boats, hauling in tons of salmon. Those helping him received a quarter of the catch, worth perhaps £14 a week (well over £1,000 a week in today's money). The fish were wrapped in bulrushes and sent to London by train. So abundant were salmon during the season that local lads on indentured apprenticeships had it written into their contracts that they wouldn't be fed salmon more than twice a week.

The same had been written into the contracts of indentured servants arriving at the East Coast colony of Massachusetts 200 years earlier. Salmon were even more abundant in the rivers there – so many of them, it was said, that during their migration upstream, you could walk across a river on the backs of salmon without getting your feet wet. And these were big fish, on average twice the size of the tiddlers we're familiar with today. Settlers along America's East Coast also found equally astounding numbers of even bigger monsters lurking beneath the water: Atlantic sturgeon.

These giant fish reached fifteen feet in length and weighed over 750 pounds. American Indian fishermen in Chesapeake Bay caught them by slipping a noose of rope over their tails – and then hanging on. A really big fish frequently pulled the fisherman out of his boat and often beneath the surface but, if he could hold on for long enough, the fish tired and could be hauled ashore. The Chesapeake settlers soon began netting sturgeon as a commercial catch to be pickled, then sent back for sale in England. In one day, along a two-mile stretch of coast, the Englishmen caught 600 sturgeons. Further north, in New England, sturgeon were so abundant they were a serious hazard for small boats plying the rivers.

Overfishing soon relegated these mind-boggling sights to history – a story that had already unfolded in Britain many centuries earlier. Go back far enough and monstrous sturgeon must have swum in similarly spectacular numbers in the Tees, where, like salmon and sea trout, they would have swum upstream from estuarine and coastal feeding grounds to spawn in gravel beds along the middle reaches of the river.* Nearly 1,500 years ago, Anglo-Saxons began trapping these huge fish on their migrations upriver using weirs like those that gave Yarm its name, and sturgeon numbers began to decline.

Even so, these huge fish were still numerous enough in the twelfth century to feed whole armies. King Stephen, the son of Henry I – he who'd died of a surfeit of lampreys – seemed set on inflicting a similar fate on his army of 80,000 men by feeding them on the plentiful sturgeon. It seems beyond belief now, but there were more than enough sturgeon in the Thames to feed his army while it was based in London. Two centuries later, in 1324, King Edward II declared the sturgeon a royal fish, which

---

* The species in British rivers is the European sturgeon, *Acipenser sturio*. On America's East Coast, two different species occur, the Atlantic sturgeon, *A. oxyrhynchus*, and the shortnose sturgeon, *A. brevirostrum*, although a few Atlantic sturgeon have been recorded in British rivers.

meant that any caught in Britain belonged to the Crown. Yet, despite this exalted status, over the centuries sturgeons had to endure not only overfishing but increasing pollution in rivers and estuaries. They're extremely sensitive both to low oxygen levels and to toxic chemicals but, surprisingly, they never quite vanished from our rivers.

However, catching one of these river monsters became unusual enough that over the last few hundred years each one hauled ashore was sure to generate excited newspaper reports. Since the advent of newspapers around 1700, 1,400 sturgeons have been recorded from rivers across the whole of Britain, nowhere near enough to feed even a small army, although some of these were still enormous fish. The largest recorded was a 400-pound beast caught in 1903 on the River Severn. However, during the whole of the twentieth century, only thirteen sturgeons were reported.

In the late nineteenth century, sturgeon still swam up the Tees as far as Yarm. We know this because George Goldie caught one, an eight-foot specimen with which he was proudly photographed. How on Earth had this fish made it through the estuary and the lower reaches of the Tees, by now lined with iron and chemical works? Many of the fish caught in Britain more recently (and this is probably also the case for older records) are females packed with eggs so, against all odds, they're clearly trying to spawn. There are, however, no records of sturgeon spawning in any British rivers, nor in most of those in Western Europe; the only place where they still spawn is in the Gironde on France's Atlantic coast. However, a quirk of sturgeon behaviour leaves conservationists with a glimmer of hope.

Like salmon, most sturgeon return to the rivers of their birth, but recent research has shown that about one in ten are more adventurous and set out to seek new rivers. This has encouraged several organisations to come together as the Sturgeon Alliance, working to improve rivers across the country to a point where these more intrepid sturgeons can re-establish breeding populations. Unfortunately, sturgeon are nowhere near as athletic as salmon, so barriers like the Tees Barrage present an insurmountable problem.

As we've seen, the Tees once brimmed with fish of many kinds, supporting fisheries in the estuary and throughout the lower and middle reaches. By the start of the twentieth century, however, all of these had collapsed entirely. In 1901, the government set up a Royal Commission to investigate the dramatic drop in fish numbers on the Tees. It was hardly rocket science. One sniff of the river anywhere from Stockton down to Middlesbrough would have provided the answer. Despite setbacks, such as the recent mass deaths of crustaceans around the estuary (see Chapter 2), the river is in better shape than it was, and many kinds of fish are more abundant. We might celebrate this, but now that we know how far our baselines have shifted, we should at the same time recognise that this is merely a tiny step forward after many centuries of giant leaps backward. The natural world of the past, when Yarm's fish weirs were in full swing – despite already suffering the effects of humanity over millennia – was still far more bounteous than today.

At the time of the Domesday Book, Yarm belonged to the king, but under William the Conqueror's son, King Henry I, Yarm passed to Robert de Brus of Skelton, and the town remained in the hands of the de Brus family for several centuries. In 1207, King John granted Peter de Brus the right to hold a weekly market in Yarm, along with two annual fairs. In the subsequent centuries, Yarm prospered – so much so that during the early years of the fourteenth century, in the kind of quirk that makes history entertaining, the wealthy town was raided by Robert the Bruce, descendent of Robert de Brus, when, as king of Scotland, the Bruce made regular sorties to settlements along the Tees and further south.

In 1674, under King Charles II, Yarm's annual fairs doubled to four a year, but by 1867, only one of those fairs survived.

It still does today.* Around mid-October, the high street is closed off for three days of revelry which these days include the obligatory Ferris wheel and fast-food outlets. However, on Saturday, there's a glimpse into a more distant past when the 'Riding of the Fair' takes place. The 'Flashing of the Horses', as it's also called, sees a motley assortment of equines raced down the high street in an echo of the days when the fair was an important market for livestock. Horses of all kinds are paraded through the street against the backdrop of the Yarm Viaduct, which strides over the town, carrying the railway line from Middlesbrough to York. It's a reminder that before the railways, horses connected the country.

Along with horses, Britain's waterways were a critical part of the nation's communication and transportation network, and remained so even after the Stockton and Darlington Railway sparked a transport revolution. Railways carried ever-larger quantities of raw materials and manufactured goods to river ports, which grew as quickly as the railways, as hubs for the increasing imports and exports. The volume of goods on the move rose rapidly through the Industrial Revolution and ports had to cope with this increased traffic. As we've seen, the port of Stockton relinquished a lot of its trade to Middlesbrough further downstream as ships became larger to cope with the demands of increased trade. But before that, Stockton itself had eclipsed Yarm as a port.

In the twelfth century, Yarm became the first port on the Tees and remained the major port on the river for 600 years, since the long loop of the river that almost surrounds the town created plenty of room for extensive, noisy, bustling quays. In contrast, the riverside here today is quiet. Follow the path past Yarm's Sainsbury's and across the car park and you'll find small jetties that overlook the river, where most of the noise

---

* Yarm fair was cancelled in 2020 owing to the Covid-19 pandemic – the first time it hadn't been run in its more-than-800-year history.

and bustle comes from flocks of black-headed gulls looking for scraps dropped by people enjoying a peaceful lunch by the river. Old pictures show a very different scene of crowded, busy quays lined with sailing ships. Vessels ranging from sixty to one hundred tons were loaded with flour, wool, grain, hides, salt and lead, bound for ports around Britain and across the North Sea.

Yarm also had a bridge, the lowest on the river before the construction of the eighteenth-century bridge at Stockton. Yarm's bridge connected two different worlds: to the south of the river Norman Britain, controlled from London, and to the north the land of the prince bishops, controlled from Durham. Standing at this strategic point on the Tees, Yarm grew to be an important communication and trade hub. The stone bridge that still spans the river here was commissioned by Bishop Walter Skirlaw in 1400, but there are references to a bridge here, probably made of wood, from half a century earlier. In 1803, traffic crossing the old bridge grew too heavy, so a new bridge, more fittingly made of iron, was built. In 1805, the mayor of Stockton duly turned up to open this new crossing with the words: 'May the Almighty protect this undertaking and may this bridge stand the test of time.' The Almighty must have been looking the other way. The bridge fell down a year later. Luckily, no one had bothered to dismantle the old stone bridge – score one for Bishop Skirlaw, who clearly did have the ear of the Almighty. A century later, in 1908, George Goldie rescued a bridge girder from the riverbed and incorporated it into the Blue Bell Inn – for what purpose, I've yet to fathom.

As a major port, Yarm also built ships, although only small ones. Yarm lay nearly twenty miles inland as the river flows but, until the barrage was built, it was still tidal. So, ships had to ride four successive high tides to reach the river-mouth. There was a limit to the size of vessel that could navigate the winding course of the river to and from Yarm, a problem that eventually doomed its port economy. The cuts made to straighten the river in the nineteenth century, across the necks of two long mean-ders downstream from Stockton, left Stockton just four miles by river from the sea and made it a far more lucrative port for both trade and shipbuilding.

The river created Yarm, but on several occasions, it almost destroyed it. Yarm is built on low-lying land, almost entirely surrounded by water, and the Tees was notorious for rapid and often unexpected rises in water level – so-called Tees rolls, when the river may rise six feet or more in less than an hour. Yarm, along with other riverside towns, was inundated by the Tees on several occasions, although Yarm's location made it especially prone to deep flooding. The Tees flowed through Yarm's streets in 1753, 1783 and 1815, but the worst disaster happened on 16 and 17 November 1771.

Torrential rain, after falling for three days and nights on the high dales where the Tees is born, flowed in sheets across the fells and into the river. The deluge surged downstream, rais-ing the river in its lower course to unprecedented levels with-out any warning. As it happened, on this fateful day, a rector from Yarm, Reverend Bramwell, was at Hurworth, overland more than six miles upstream from Yarm, but twice that along the winding river. He saw the waters rising rapidly and knew what this would mean for Yarm, so he dispatched a messenger to ride with all haste carrying a warning to the townsfolk to move anything they could to the upper floors of buildings, and then to move themselves and their livestock to higher ground. When the messenger arrived in Yarm, it had been raining heav-ily since the early morning, but the Tees was still flowing peace-fully around the town. With no sign of an impending cataclysm, the people of Yarm chose to ignore the warning. The locals thought they knew the Tees in all its moods, but in 1771 it surprised everyone. The water rose faster than anyone had ever seen before and quickly overwhelmed the town.

The river rose so high, in fact, that it flowed over the para-pets of the old stone bridge which, nevertheless, stood firm against the flood, as it must have done many times over the centuries. However, the bridge's arches deflected some of the water, sending it as a torrent down the streets of the western side of the town and sweeping away all before it. In almost no time, parts of the high street were under twenty feet of water and nine people had drowned. Every last house in town was underwater.

The great November storm of 1771 flooded towns right along the Tees. In 1818, William Garrett collated first-hand accounts of this disaster into a report which, by detailing dramatic stories of individual people caught up in the flood, makes the whole episode feel very immediate, even after more than two centuries. At Stockton, the flood was 'greater than ever known in the memory of man'.* Surprisingly, however, apart from inundating one warehouse and three liquor cellars, the flood did little damage here. But the people of Stockton knew Yarm would be a lot worse-off. 'As soon as the flood was discovered at Stockton, carriages were procured to take two boats over land to Yarm, with some experienced sailors, who, thereby, saved the lives of many.' Elsewhere, Croft and Barnard Castle were flooded, and the Tees was half a mile wide at Low Coniscliffe near Darlington. At Gainford, part of the churchyard was washed away and coffins were carried off downstream.

The three rivers of the North-East, the Tees, Wear and Tyne,† all rise within a few miles of each other in the Pennines and the storm itself spread widely across the lowlands, so all three rivers rose in devastating floods. Along the Tyne, all but one of the bridges were washed away. In Durham, on the Wear, more than thirty ships were wrecked. The floods were so severe

---

* William Garrett. *An Account of the Great Floods in the Rivers Tyne, Tees, Wear, Eden, &c. in 1771 and 1815*. E. Charney. 1818.
† The Tyne is fed from two sources, the North Tyne, rising in Northumberland, and the South Tyne, which rises just a couple of miles from the source of the Tees.

that the events of November 1771 became known as the Great Inundation. Some reports suggest that the water may have risen to twice the highest level ever recorded – before or since. So many bridges were destroyed that the authorities were forced to react by appointing more highly qualified engineers and architects to design and build replacement bridges and by instigating more methodical and regular bridge inspections to make sure they would stand up to future floods. They obviously missed Yarm's ill-fated iron bridge.

Walking along the riverside path on a sunny day, it's hard to imagine that the river has such a dark side, but the evidence is still there. In several places around town, the high-water mark of the 1771 flood is indicated on the walls of buildings. Looping back to the high street from the river brings me to the Town Hall, on which the flood height has also been marked – well above my head – a powerful reminder of the destructive force of the Tees.

Heading back north along the high street to cross back over the old stone bridge and rejoin the Teesdale Way, I pass the George and Dragon, and a brief refreshment stop here connects us with another important part of our story. This is the pub in which an influential group of Darlington and Yarm men met to promote the idea of a railway to link the Durham coalfields with the port of Stockton, which was promoting an alternative plan to improve transport links as demand for coal soared – but Stockton's plan was based on a canal system. Perhaps still harbouring a grudge against Stockton for supplanting Yarm as the most important port on the Tees, the Yarm men, along with the scheme's main promoter, Darlington-based Quaker Edward Pease, were instead firmly in favour of developing the world's first public railway system. That meeting, a seminal moment in Britain's industrial history, is still commemorated by a plaque on the pub wall.

Stepping back out into bright spring sunshine, there's an impressive monument to the prescience of those men gathered in the pub in 1820. The forty-three arches of the imposing Yarm Viaduct carry the railway line high over the old town. The viaduct was built just a couple of decades after the first locomotives chugged along the original Stockton and Darlington Railway, when the Leeds and Thirsk Railway extended its lines through Northallerton as far as Stockton and Hartlepool. It's a striking reminder of the speed at which rail networks spread across the country in the mid-nineteenth century. The viaduct steps across the river right next to Yarm Bridge, more than six hundred years of Yarm's pivotal role in British history encompassed in one view.

Leaving Yarm, the Teesdale Way hugs the north bank of the winding river along the southbound arm of the Yarm meander, where it runs parallel to the Yarm Viaduct, providing a chance to appreciate the scale of the engineering that was transforming the Tees in the nineteenth century. After less than a mile, the river swings westwards and in another half-mile, I pass the settlement of Aislaby on my right, noteworthy for being one of the few Viking place names north of the river, although, as we'll find out shortly, the Vikings did establish some substantial holdings to the north of the Tees. It's notable in my mind for two other reasons. It was the home of Steve McClaren, the most successful manager of the Boro in many a year.[*] It also featured in one of the prophetic utterances of one Ursula Shipton – Mother Shipton – the famous 'witch' who lived five hundred years ago in Knaresborough, near Harrogate, thirty miles to the south of Aislaby.

With the usual obtuse language of such prophecies, Mother Shipton, Yorkshire's very own Nostradamus, claimed that, 'When Yarm sinks and Egglescliffe swims, Aislaby will be the market town.' I've no idea what this means. Yarm certainly

---

[*] Under McClaren, Boro won the League Cup in 2004, our first major trophy, and he took them to the UEFA Cup final and the FA Cup semi-finals in 2006. As near to the glory days as we've ever been.

sank back in 1771, but Aislaby remains a small though rather well-to-do village. Nor, on searching through her other prophecies, did she seem to have had anything to say about when Middlesbrough FC will return to the success they had under Steve McClaren.

Ursula also gave her name to the day-flying moth, the Mother Shipton, which I found in several grassy spots earlier in my journey. Mother Shipton was said to have possessed the classic face of a fairy-tale witch, with a long, crooked nose and pointed chin, and on each forewing of the eponymous moth there's a passable likeness in profile of a hag's face. It's an odd, if apt, name for this moth, but it's only one of many hundreds of similarly evocative moth names whose derivation and history are as fascinating as the moths themselves.

I use a light trap in my garden to catch moths for photography or for breeding, but I also look forward to emptying the trap each morning just to see what strangely named creatures are lurking inside – each one a history lesson from an era when naturalists clearly had a lot more imagination. In no other group of animals can you come across a conformist and a non-conformist or an argent and sable, another day-flying moth named after the heraldic terms for the black-and-silver markings that pepper its wings. Or how about the peach blossom, the Saxon, the maiden's blush, the sprawler, the Merveille du Jour or the lunar spotted pinion?* I barely notice that I've left Aislaby behind as I recite in my head all the moth names I can remember, like some arcane litany – or perhaps a Mother Shipton spell.

In a mile or so, I find myself looking across at Low Worsall on the south bank of the river. This is yet another noteworthy point on this part of the journey because, before the Tees

---

* Should you find yourself becoming as fascinated as me by the origins of such fantastic names, then I recommend Peter Marren. *Emperors, Admirals and Chimney-Sweepers: The Weird and Wonderful Names of British Butterflies and Moths*. Little Toller Books. 2019.

Barrage was built, this was the highest tidal point on the river, some twenty miles from the lighthouse at South Gare and the open sea beyond. It seems a long time ago that we explored the slaggy grasslands, dunes and open mud around the Gares, and yet the sea's influence would, in the past, have been with us all the way to this quiet rural spot.

Since the river remained tidal this far into the heart of the Tees Valley, it could still, in theory, carry ships laden with local produce downriver, and, in the early eighteenth century, Thomas Peirse tried to turn theory into practice. In 1732, he sank an inherited fortune into building Peirseburgh, the highest port on the Tees. He built stone quays, granaries and warehouses at Low Worsall and, as his ambitions and the scale of the enterprise grew, houses for dockworkers and a pub where they could slake their thirst. That pub, the aptly named Ship Inn, is still there – and still slaking thirst – though its name seems oddly out of place now along a tranquil stretch of river flowing through verdant fields and scattered copses.

Thomas's first ship was a sloop, built with a cunning folding mast so he could float under the bridge at Yarm, no doubt to the chagrin of the townsfolk, since he was planning to steal their trade. By 1750, Thomas had more ships, carrying Norwegian timber and tar upriver to Low Worsall and returning downstream loaded with local corn, cheese and ham. A big part of his trade, however, was lead from the mines in Swaledale to the south. His selling point was that Peirseburgh lay at the end of a shorter, easier journey for heavily laden pack horses. He had a point. At the time, the roads into Yarm were very poor and Low Worsall did lie a little closer than Yarm to the areas of lead production.

The local producers obviously agreed and for a generation the little port prospered, although that might have been more to do with the illicit activities around Low Worsall than with the busy quaysides. History paints Thomas as a bit of a scoundrel on several counts. There were always rumours of smuggling, but in 1958 a tunnel was discovered that ran from the river-bank up to the hall that Thomas had built for himself. In 1961, another tunnel was uncovered, this time leading from the river to the Ship Inn. When they were in town, local customs officials stayed in the Maltkiln Inn (now Peirseburgh Grange), which is well away from the places where imported goods were being spirited away from the riverside.

Nor did Thomas endear himself to the inhabitants of Yarm downstream. Despite assistance from the falling tides, it was still a difficult and lengthy journey from Low Worsall along the winding Tees to the North Sea. Then, as now, time was money, so Thomas used gangs of men on the riverbank to drag his ships downstream. To clear their way, they broke down fences and gates and trampled riverside paths into a quagmire right around the perimeter of Yarm. The gangs of men were a tough lot, not to be trifled with, but eventually the Yarm locals had had enough. They took a case against Thomas to York Assizes – and won. The court ruled that, rather than 'haling' his ships, as Thomas had been doing, a series of stout poles should be erected along the riverbank. A rope tied to a pole and passed to the ship then allowed men on board to pull the ship towards each pole in turn.* Thomas, however, simply ignored the court.

He further angered Yarm residents when he dug out gravel beds to deepen the river. In doing so, he destroyed spawning grounds of the salmon which supported a valuable fishery in Yarm. Yarm retaliated by charging a toll for the use of the roads into the town, which thus became turnpikes and provided their owners with the finance to keep the roads in better order.

---

* This process is called *warping* the ship. The same process can be carried out at sea by a ship's boat carrying the anchor far ahead of the bow, then dropping it. Sailors can then use the capstan to wind the ship towards the anchor, repeating the exhausting process as often as needed. This is the way that Captain Cook extricated himself from the treacherous Great Barrier Reef in Queensland, and so – yes – Captain Cook had warp drive long before Captain Kirk.

Meanwhile, long before that fateful meeting in the George and Dragon, there were widespread discussions about canal networks or rail systems to improve local transport, all of which would make Peirseburgh redundant. In addition, the Great Inundation of 1771 must have done considerable damage to Peirseburgh.

When Thomas Peirse died in 1779, his son discovered that the debts and liabilities of the company he had inherited far outstripped its value, and he was immediately declared bankrupt. The quays, though, continued to be used right up until 1820, just five years before the Stockton and Darlington Railway revolutionised the whole lower Tees Valley and far beyond. The quays are no longer visible, although stones from their construction were used to build the All Saints church in Low Worsall 1893, leaving only the Ship Inn and a couple of smuggling tunnels as reminders of a much busier place.

Along this stretch, the river is sometimes flanked by steep slopes, often covered in woodland since the land is too inaccessible to cultivate. These wooded banks lie around the outside curve of each of the many meanders, where the flow is faster, and the river is cutting back into the landscape as the meander slowly grows longer. The steep banks also shape some of the villages along the river. After a few more meanders, which the Teesdale Way thankfully sidesteps by taking the shorter route across the meander neck, I arrive at Middleton One Row, named, logically enough, because the main street faces the river and so has a row of houses along one side only. On the other side there's a sloping village green, a little too steep for a good game of football, which leads to an even steeper strip of woodland that runs down to the river.

These meander woodlands are one of the joys along this stretch of the river. In places, the Teesdale Way closely follows the riverbank, passing through woodlands in which, just occasionally, I meet other walkers. I could be miles from anywhere as I sit on the riverbank at Middleton One Row, taking in the rural landscapes on the opposite bank. Yet, I'm only a few hundred yards from the end of the runway of Teesside International Airport. I'm also close to another revolutionary communication network, although one which predates air and rail travel by almost two millennia.

A Roman road, called Cade's Road after a local historian, crossed the Tees here over a long-vanished stone bridge, still remembered in Pountey's Lane in Middleton One Row, a name derived from Pons Tees, from the Latin *pons* for 'bridge' – hence Pons Tees or Tees Bridge. From the Tees, the Roman road headed north from Middleton One Row through Sadberge on its way to the Tyne and a Roman port at present-day South Shields.* There was also a Roman fort at Sadberge, around three miles north of the Tees, built on the highest point of land in the area. Indeed, Sadberge means 'flat-topped hill', but not in the Latin of the Romans; around three centuries after the legions left Britain, the Vikings recognised the value of this commanding position.

It's worth making a brief detour to Sadberge, although, since it lies between the busy A67 and A66 roads, I decide to drive rather than walk. Approaching the village, the eponymous hill is barely noticeable. It's not so much a hill as a gentle rise in elevation, yet Sadberge is still high enough to give me an overview both of our journey so far and of what is yet to come. On a clear day, through gaps between the buildings in this quiet village, you can see twenty miles to the east, as far as Redcar and Hartlepool, straddling the mouth of the Tees. Turn the other way and the area around Barnard Castle is just about visible twenty miles to the west, marking the start of the high country

---

* Cade's Road (its Roman name is unknown) ran for 100 miles, linking the Humber with the Tyne. It passed through Thornton-le-Street near Thirsk in Yorkshire and Chester-le-Street in County Durham – their names signifying that the original settlements lay near a Roman road.

of the dales. Although this section of our journey seems to be turning into a pub crawl, I would plead in my defence that the Three Tuns in Sadberge – close to the site of the old Roman road at the top of the strategically located flat-topped hill – is the perfect place to reflect on the contorted post-Roman history of this obscure part of Britain.

We saw in the last chapter that, from the eighth century onwards, the most extensive Viking settlements lay to the south of the river. Sadberge's prominent hill, however, became the centre of an important Viking settlement to the north, the Wapentake of Sadberge. A *wapentake* was a Viking administrative district, and the boundaries of such districts often outlived their Viking creators by many centuries. Part of the Wapentake of Langbaurgh, which in Viking times ran to the south of the Tees from Low Worsall to South Gare, and which therefore covered a substantial area of the north-eastern corner of the North Riding of Yorkshire, was partially resurrected in 1974, in countrywide boundary changes, as the district of Langbaurgh, to the south of Middlesbrough.

The Wapentake of Sadberge occupied a similar large area to the north of the river, running from the low dales around what would later become Barnard Castle and Newbiggin to Teesmouth and the coast at Hartlepool. *Wapentake*, incidentally, means 'weapon taking' or 'weapon touching' and refers to an oath of loyalty sworn to an overlord by all the leading men living in the district. The oath was taken when they all gathered, perhaps on the hill at Sadberge, to clank their swords together. Sadberge as a distinct district, however, also outlasted the Vikings.

In the ninth century, the old Anglo-Saxon kingdom of Bernicia (covering much of northern England, with the Tees as its southern boundary) began to dissolve into two areas that would become roughly the modern counties of Northumberland and Durham, divided by the Tyne. But, for some reason, Sadberge – despite being nearly thirty miles to the south of the Tyne

– became part of Northumberland, under the control of the earl of Northumberland. In practice, though, its location, remote from the earl's seat of power, meant that the Viking district continued as a relatively independent county. It wasn't until 1075 that William the Conqueror seized Sadberge, in retaliation for the earl of Northumberland siding with the Scots against the invading Normans. In this way, it remained out of the hands of the increasingly powerful prince bishops of Durham as they gained their own independent control of the County Palatine of Durham – although, not for long.

In the twelfth century, crusading was a popular, if expensive, pursuit. The most famous crusader was King Richard I, Coeur de Lion, who spent far more time in the Near East or France than he ever did in England. There were, however, plenty of other people around the country just as keen to join in, including Prince Bishop Hugh Pudsey. Pudsey was a nephew of King Stephen and the most princely of the prince bishops, with a taste for the flamboyant. He levied heavy taxes on the citizenry of Durham to raise money for a ship to carry him in appropriate splendour to the Holy Land.

When King Richard saw the vessel, complete with silver throne, he decided that he didn't want to be outshone by a bishop – even a prince bishop. He suggested that Pudsey might prefer to stay at home to look after England during Richard's long absences. Pudsey agreed, so long as Richard made him earl of Northumberland *and* earl of Sadberge. But money would also have to change hands. For the price of £11,000 (worth over £17 million today, according to the Bank of England's handy inflation calculator), Pudsey (and the prince bishops that followed him) acquired the last missing piece in the jigsaw of County Durham, along with the rights to tax its inhabitants and hold courts.

In later years, courts were held in what is now the Three Tuns pub, where the beer cellar was originally the dungeon. Sadberge's local tribunals and assizes lasted for centuries and kept a semblance of the independence that the area had enjoyed during the Middle Ages. The sheriffs of Sadberge presided over a county court here up until the seventeenth century, and when their jurisdiction finally ended, the building was converted into the Three Tuns pub. Yet the idea of Sadberge as an entirely separate and distinct part of England has been hard to extinguish. In 1838, the earl of Northumberland and Sadberge was a distant memory, but during the reign of Queen Victoria, the people of Sadberge proclaimed her sovereignty over the old independent *wapentake*. There's a stone on the village green in Sadberge commemorating Victoria's Golden Jubilee, which reads: 'This stone was placed here to commemorate the Golden Jubilee of Victoria, Queen of the United Kingdom, Empress of India, and Countess of Sadberge 1867.'

Immediately to the south of Sadberge, the Tees throws itself into one of its longest meanders, enclosing the hamlet of Sockburn. This area, too, seems to have been an important one for the Vikings. Several splendid hogback stones have been found within the Sockburn meander, nine of which are held in the chapel of the now-ruined All Saints Church, close to Sockburn Hall. These stones seem unique to this area. They're about six feet long, shaped a little like a Viking longhouse and covered in often-elaborate carvings. They may have been used as grave markers, possibly by Christianised Vikings, during the tenth century. If so, the land enclosed by the Sockburn meander was probably an important religious centre in the Wapentake of Sadberge.

Across the river, in the Wapentake of Langbaurgh, the same seems to have been true of Roseberry Topping, the distinctive hill on the northern edge of the North York Moors near Great Ayton, which is often linked with the Norse god Odin, father of all the gods. Incidentally, apart from the hogback stones, a carving on a stone from an ancient church in Kirklevington, a little to the south of Yarm, is reputed to portray Odin with his two ravens, Huginn and Muninn – Thought and Memory* – perched one on each shoulder. This evocative carving has been dated to around AD 900 and today is on display next to the Yarm Helmet at Preston Park Museum. Taken together, these artefacts make the Vikings in this corner of the North-East feel very real. The strong Viking presence here may also explain an enduring myth about this stretch of the river.

We've already encountered some very real monsters in this part of our journey, in the shape of large fish such as sturgeons. But the river at Sockburn was home to an even larger and far more fearsome beast – the Sockburn Worm. If a 'worm' doesn't sound particularly intimidating to our modern ears, consider that in the past this was a general term applied to all manner of dragons and wyverns. Any decent medieval field guide would quickly point out the differences between these two formidable beasts of legend. Both have large, leathery wings, but whereas dragons also possess four legs, wyverns have but two. The evolution of flight in dragons therefore calls to mind (at least to my mind) the different origins of flight in vertebrates and insects. When vertebrates (bats, birds and pterosaurs) conquered the air, they had to turn their front legs into wings, leaving them two-legged. Insects, on the other hand, invented wings from scratch, leaving all six of their legs to carry on working as legs. Thus it must have been with the evolution of dragons and wyverns.

The Sockburn Worm, it turns out, was a wyvern – and a particularly fierce one – doing what wyverns do, devouring local people and generally causing mayhem. Cue the inevitable dragon-slayer. In this case, it was a local knight, called John Conyer, who took on the task. He slew the monster with

---

* Huginn derives from the Old Norse *hugr*, meaning thought. Muninn, from *munr*, is harder to translate but has connotations of desire, emotion and thought – so the usual Anglicised name of one of Odin's ravens, Memory, is not very accurate.

blows from his falchion, a kind of medieval sword that looked (and clearly worked) a bit like a meat cleaver. Conyer certainly cleaved the Sockburn monster and, so it's said, in return for this heroic deed his family received substantial lands around Sockburn and Yarm. The worm is long gone, but Conyer's falchion still survives. Today it's housed in Durham Cathedral, appropriately enough, because not long after Conyer's valiant efforts, a tradition was inaugurated whereby the sword was used in a ceremony to welcome each new bishop of Durham to the bishopric.

Traditionally, the new incumbent stepped into County Durham by crossing a ford near Neasham, at the upstream end of the Sockburn meander. As he did so, he was presented with Conyer's falchion by the lord of Sockburn Manor with the words:

> My lord bishop, I hereby present you with the falchion wherewith the champion Conyers slew the worm, dragon or fiery serpent which destroyed man, woman and child; in memory of which the king then reigning gave him the manor of Sockburn, to hold by this tenure, that upon first entrance of every bishop into the county the falchion should be presented.

Instead of getting their feet wet, the bishops later crossed into Durham over the bridge at Croft-on-Tees, the town in which we'll shortly arrive to conclude this particular leg of our journey. A bridge was constructed at Croft in 1400, at the same time as the bridge in Yarm, instigated once again by the bridge-building Bishop Walter Skirlaw, who clearly saw the advantage of linking his Durham hegemony to the rest of Britain. With a few lapses, the ceremony of crossing the Tees has continued up to the present day. The last of the prince bishops, William Van Mildert (reigned 1826–36), was presented with the sword in 1826. He was the last to hold the vestiges of that former princely power, which he wisely used to found Durham University.*

The Sockburn Worm is just one of several worm legends around the North-East. The Laidly Worm hung out near Bamburgh, feasting on the good folk of Spindleston Heugh (apparently it was the cursed daughter of the king of Bamburgh). More famous still, the Lambton Worm swam in the River Wear. This one has an intriguing kernel of natural-history truth about its legendary origins. It reputedly grew from a small eel-like creature, caught by one John Lambton, which had nine holes along the side of a head that resembled that of a salamander. To my mind this is a fairly accurate description of the lampreys that we met along the River Leven. Thrown down a well, Lambton's lamprey grew into a voracious monster, the moral being, presumably, that one should not drop lampreys into wells, or – in the case of King Henry I – eat too many.

Some historians, however, have postulated another explanation for these monsters, based on the history of the North-East. It's been suggested that the medieval worms were born from ancient memories of Viking invasions. Their longships carried carvings of worms (*ormr* in Old Norse) on their bows. However, Viking ships were a familiar sight across large parts of the country, whereas worm legends are much more restricted, so I'm not convinced by this explanation. I prefer the idea of super-sized killer lampreys. However, Ormesby, the part of Middlesbrough in which I went to school, derives from the same word – *Ormr's by*, 'Ormr's village'. I don't know who

---

* The ceremony was held again in 1860, when the third of the modern bishops, Henry Montagu Villiers (bishop for the brief period of 1860 until his death in 1861), entered the county – this time, appropriately enough, by train. He had the train stopped halfway across the rail bridge over the Tees to receive the sword. The ceremony was revived again in 1984, to welcome Dr David Jenkins (bishop of Durham 1984–94), although on this occasion the traditional speech was given by the mayor of Darlington.

Ormr was, but if he was named after a dragon, I'm not sure I would have wanted to find out.

The Sockburn peninsula is a remote spot today and difficult to access, although Sockburn Hall, home of the Conyers family, still exists. However, at the very end of the eighteenth century, this quiet, scenic location drew two of our best-known poets. In 1799, William Wordsworth (1770–1850) courted Mary Hutchinson here, before eventually setting off, with his sister Dorothy, to cross the Pennines to live in the Lake District. Samuel Taylor Coleridge (1772–1834) also stayed at Sockburn, where, even though he was already married, he fell in love with Mary's sister, Sara. It was here that Coleridge wrote the poem 'Love', dedicated to Sara, in which he describes a knight with a burning brand on his shield, very possibly the mailed knight in the Conyers' tomb beneath the ruined Sockburn church. A year before visiting Sockburn, Wordsworth and Coleridge had already collaborated on *Lyrical Ballads* (1798), which helped launch the English Romantic movement, whose philosophy encompassed some early modern environmental thinking such as the value of nature not just for profit but for the human spirit. We'll be returning to the Romantics in the next chapter.

Instead of following the long loop of the river around Sockburn, the Teesdale Way cuts a path across the neck of the meander, joining the road into Neasham. After some rough and almost impassable stretches of track outside of Yarm, the Teesdale Way here is a pleasant, gentle stroll along the open riverbank on a well-maintained track, along a high flood bank above the riverfront village green. So far, the river has flowed deep and silent beside the Teesdale Way. Now, approaching Neasham, I hear the burble of water rippling over shallows. Ahead of me, a grey heron stands sentinel in the middle of the river, in no more than a few inches of water, intently focused on procuring its next meal.

There are two ancient fording points across these shallows at Neasham, although it seems that it was the upper High

Ford that was used by the paddling prince bishops.* The path first passes Low Ford, where an ancient-looking dirt path dips into the swirling, bubbling waters of the Tees. Even though the worm has been slain, the crossing over loose stones, avoiding the deeper channels, still looks treacherous, and the decision to move the welcoming ceremony for newly inaugurated bishops of Durham to the safer and drier crossing at Croft Bridge seems eminently sensible. High Ford has now disappeared, at least from the Neasham bank, swallowed up in major flood-defence works along Kent Beck, a small tributary of the Tees. At the site of High Ford, the Teesdale Way veers away from the river, north along Kent Beck until it joins the road through Hurworth Place and Hurworth before crossing Skirlaw's bridge into Croft-on-Tees. Since no swords are on offer, I cross over to visit the twelfth-century St Peter's Church, which stands right next to the bridge – a place where in the nineteenth century, the Sockburn Worm seems to have been resurrected.

In 1843, Reverend Charles Dodgson moved from Cheshire with his young family to become rector at St Peter's. His eleven-year-old son, Charles Lutwidge Dodgson, must have soaked up the stories he heard along this stretch of the Tees, including tales of the Sockburn Worm and its slaying by John Conyers, because young Charles soon developed a vivid imagination, along with a talent for composing nonsense verse. Although he later moved to Oxford, he always regarded the rectory at Croft as home and several places around Croft crept into his work as he began to write the stories that still fire the imaginations of children today. Rather than writing under his own name, he chose a pen name. We know him today as Lewis Carroll.

He wrote the first verse of one of his most famous nonsense poems in 1855, while at Croft. It concerns a fearsome

* I've also heard it suggested that the original ceremony may have taken place at a similar ford on the Sockburn meander.

monster known as the Jabberwock and the complete poem, 'Jabberwocky', including the eventual demise of the monster, slain by a 'vorpal blade', appeared nearly twenty years later in *Alice Through the Looking Glass*. The poem is, as Alice also found, 'rather hard to understand', but, as far as can be deduced from the nonsense words, there are strong similarities to the story of Conyers and the Sockburn Worm. However, I feel compelled, as a pedantic biologist, to point out that John Tenniel's illustration in *Alice Through the Looking Glass* is of a dragon, not a wyvern.

Carroll found yet more inspiration inside St Peter's itself. Around the back of the church, where the Teesdale Way runs along the edge of the old cemetery, I meet with Paul Dunn, who is busy placing an information sign next to the grave of Dodgson Senior, ready for an upcoming open-day event. Paul is a gifted storyteller and knows the history of the church and of the local area in intimate detail – and a very peculiar history it is. First, the church itself *is* 'peculiar' – not part of the local diocese but under royal control – and there are very few such churches left, most having been absorbed into their local dioceses in the nineteenth century. Second, as Paul shows me around the church, he regales me with tales as peculiar as anything Alice encountered in Wonderland.

He ushers me to three ancient stone seats to the side of the altar, where the clergy sat. They're surrounded by images, carved in the fourteenth century, of people and strange beasts, including the face of a cat, grinning wildly. This may well have been Carroll's inspiration for the Cheshire Cat in *Alice in Wonderland*, Paul explains. Although the expression 'to grin like a Cheshire cat' first appeared in the eighteenth century, decades before Alice's encounter, Croft's grinning cat predates all of these by many centuries. Paul feels sure that the young Charles Dodgson would have been drafted by his father as an altar boy, where, kneeling at the altar rail, he could have stared up at the smirking cat. However, as Paul demonstrates, if you rise from the kneeling position, the cat's grin disappears. Very curious. I am convinced.

It may be that other well-known elements of the 'Alice' novels were also inspired by Carroll's time around Croft-on-Tees, such as the strange, backward writing on mirrors in the rectory. The rectory garden might have been in Carroll's mind as he set the scene for Alice's first adventure, and, in 1950, when the floorboards of the Croft rectory were lifted, a white glove was found, perhaps once belonging to the Dodgson family – or perhaps to a white rabbit in a hurry. Paul also thinks that an unedifying episode in the church's long history may have been behind Carroll's inclusion in *Alice in Wonderland* of the twins Tweedledum and Tweedledee, who agree to have a battle but never get around to sorting it out.

In the fifteenth century, two wealthy families, the Clervaux of Croft Hall and the Places of Halnaby Hall, just down the road, fell out in a big way. Richard Clervaux thought nothing of poaching on Roland Place's estates and both parties happily appropriated any of the other's cattle that strayed from where they belonged. They also poached each other's best servants. Their feud extended to the parish church, where they argued endlessly over where each family should sit, even bringing weapons to the services to bolster their arguments. It took Richard, Duke of Gloucester, later King Richard III, from his base at Middleham Castle, just twelve miles to the south of Croft, to bang their heads together. He suggested the radical solution of fencing their land – and that in church they simply sit where they'd always sat, Clervaux on the south side, Place on the north. The two were so pleased with this profound insight that they jointly funded the construction of the porch through which I'd entered the church.

As strange a story as this is, the church bears yet other marks of petty rivalries and jealousies. In their allotted place to the south of the aisle, the Clervaux built their own altar, fenced off by a wooden screen from both the common folks and the Milbankes, the family who came to possess Halnaby Hall in the sixteenth century. Not to be outdone, the Milbankes built a bank of raised and enclosed pews, complete with heavy red curtains, from which they could look down on the no-doubt-bemused congregation – and right into the Clervaux private chapel. Today, two large tombs occupy the north and south sides of the church – to the south, that of Richard Clervaux, and to the north, that of Sir Mark Milbanke, a wealthy Newcastle merchant. You can't help but feel that they've been trying to stare each other out for the last three and a half centuries.

A little upstream from Croft-on-Tees – and no doubt visited by the curious young Charles Dodgson – are two water-filled holes, said to be infested with monstrous fish and the souls of the damned, known appropriately enough as Hell's Kettles. At least one geologist has suggested that these holes and the way they formed may have been the inspiration for the rabbit hole that deposited Alice in Wonderland.

Hell's Kettles lie in the middle of an ordinary-looking field right alongside the A167, a mundane setting for a place that has attracted attention since the twelfth century. They were said to have been formed during a great earthquake in 1179, when the ground rose up like a great tower, then crashed back to earth, leaving a series of deep holes which subsequently claimed the lives of many people and animals – eaten alive by the swarming shoals of giant pike and eels that were said to infest them. Locals also claimed that the pits were bottomless, although in

1958, risking ravenous eels and pike, divers reported that the pools are in fact little more than twenty feet deep.

For centuries, they drew curious visitors from all around the country, including a cynical Daniel Defoe, who dismissed them as old coal pits – not that coal has ever been mined in this part of the Tees Valley. But Defoe, yet another in the long line of literary figures to visit the Tees, did have a point. Hell's Kettles owe their formation to the underlying geology, created not in an apocalyptic earthquake but in a sudden and catastrophic subsidence.

This area is underlain by thick bands of gypsum (hydrated calcium sulphate), the main ingredient of plaster of Paris. Gypsum dissolves much faster than limestone when water percolates through it, soon forming large underground cavities. When these cavities grow too large, the surrounding rocks give way and the ground above collapses without warning to create a deep hole. This probably happened in the twelfth century, to judge from later medieval descriptions, and since then the pools have, rather curiously, developed entirely different ecologies.

Early references describe three pools here, but today two have fused together to create the Double Kettle. This larger kettle is fed by surface runoff and looks like any other small pond in the area. Surrounding the pond is a fringe of common reed and beneath the turbid water there are sparse growths of the introduced Canadian pondweed and the native fennel-leaved pondweed. The smaller Croft Kettle, on the other hand, is fed by an underground spring which has percolated through the calcareous rocks. The water is therefore clear and contains high concentrations of base minerals. This encourages the growth of fen vegetation, more like that found in the fens of East Anglia than in the rest of the Tees Valley. Croft Kettle is surrounded by stands of great fen sedge mixed with lesser water parsnip, mare's-tail and tubular water dropwort. Beneath the surface there are thick growths of two kinds of stonewort, *Chara hispida* and *C. vulgaris*. These plants look like any other pondweed, but they're not even remotely related. They're green algae that grow into complex, pondweed-like shapes, supporting themselves by secreting an external skeleton of calcium carbonate. For this reason, they're restricted to areas where the water has high concentrations of calcium – and Croft Kettle is ideal.

Could Lewis Carroll have used these 'bottomless' pits as inspiration for the rabbit hole through which Alice fell 'down, down, down'? Well, if not these kettles, close to his childhood home, then perhaps similar features around Ripon, twelve miles to the south, where his father eventually became canon. Intriguingly, the original inspiration for Alice,* Mary Badcock, lived in Ure Lodge near Ripon, where, in 1834, the ground also collapsed to create a sixty-foot-deep hole – an event that must have caused plenty of speculation.† Curiouser and curiouser.

* Photographs of Mary Badcock were used by John Tenniel, who illustrated Carroll's books, as the basis for his drawings of Alice.

† Another collapse in 1997 left a twenty-foot hole. Around Ripon, there are lots of deep holes in the ground.

7

Green and Pleasant Land

Croft-on-Tees to Rokeby – Railways, Romans and Romantics

> And did those feet in ancient time,
> Walk upon Englands mountains green:
> And was the holy Lamb of God,
> On Englands pleasant pastures seen?
> ...
> I will not cease from Mental Fight,
> Nor shall my sword sleep in my hand:
> Till we have built Jerusalem,
> In Englands green & pleasant Land.
>
> WILLIAM BLAKE, 'Jerusalem', 1804

F rom Croft-on-Tees, our journey will eventually take us through the agricultural landscapes of the Tees Valley and further back in time than we've ventured so far, but not before we explore another major tributary of the Tees. A few steps upstream from the bridge at Croft-on-Tees bring me to the confluence of the Tees and the Skerne, a river which, before it empties into the Tees, runs for twenty-five miles from its source in the Magnesian limestone hills of County Durham. Magnesian limestone is a distinctive form of limestone which outcrops in a narrow band, just a few miles wide, running north from Nottinghamshire through Yorkshire and County Durham. It creates a rare landscape that is always fascinating to explore and one that will more than repay the efforts of a short excursion along the Skerne.

Other kinds of limestone outcrop more widely across the British Isles. Older and harder limestones make up the Pennines and other western mountain regions. We'll meet these extensive limestones, laid down in the Carboniferous Period (about 360 to 299 million years ago), when we reach the upper course of the Tees. Younger, softer limestones were laid down in the Jurassic Period (about 200 to 145 million years ago), and now form the rolling hills of the Cotswolds of England's South-West.* Younger still, the chalk of the Cretaceous Period (about 145 to 66 million years ago) shapes the North and South Downs and runs in a broad spur up through eastern England. The Magnesian limestone was laid down about 250 million years ago in the Permian Period, between the older Carboniferous and the younger Jurassic and Cretaceous Periods. Magnesian limestone formed at the bottom of a rapidly evaporating desert sea during a time when the patch of Earth that would

---

* The North York Moors are also of Jurassic age, but the rocks there, laid down in a different arm of the sea from the Cotswolds, are mostly sandstone. However, the Tabular Hills on the southern edge of the North York Moors are composed of Corallian limestone, laid down in the late Jurassic.

eventually become England's North-East was an extreme version of the Sahara.

These Permian limestones have a different composition from other kinds of limestone, which are composed largely of calcium carbonate. In contrast, Magnesian limestone consists of calcium magnesium carbonate. It's a form of dolomite, useful for building heat-resistant bricks as well as being a vital additive in blast furnaces, where it reacts with impurities in the iron ore to leave pure molten iron. We've already explored the flora of the man-made limestone of slag heaps around the Tees Estuary, but the flora of the original Magnesian limestone is even more interesting.

The Skerne flows close to the village of Bishop Middleham, to the north of which is a quarry gouged into the Magnesian limestone. Part of the site is still being quarried, but the northernmost excavations have been long abandoned and reclaimed by nature. A narrow road runs out of the village, past the working quarry, after which it reaches a scrubby hedge that looks much like every other hedge along the road except that this one has a narrow gate in it, inviting you to investigate further. It's hard to spot at first, until you see the 'Durham Wildlife Trust' sign. Hop over a stile and through a patch of rough scrub and you'll find yourself in a world of wonder.

A path runs through grassland and scrub around the western rim of the old quarry, past some low, honey-coloured cliffs of limestone, until it reaches a long series of steps that descend to the short turf on the quarry floor. Every step along this path reveals something worth closer investigation. Clumps of rockrose sprawl over the ground, covered in yellow flowers as bright as the sunshine. The magenta cones of pyramidal orchids and the translucent pink spikes of fragrant orchids poke up from the grassy carpet. The darker, velvety flowers of bee orchids are harder to spot, but once you get your eye in, they seem to be everywhere. Tiny flecks of white, barely visible among the grass stems, are the minute flowers of fairy flax. But the plant that steals the show here is another orchid – the dark-red helleborine.

This orchid grows in just a few locations around the British Isles, often on the patchy, shallow soils that accumulate on otherwise-bare limestone pavements or even directly out of tiny cracks in the rocks themselves. Scattered plants grow on the spectacular limestone sheets that run down to the wild Atlantic in the Burren of Co. Clare in in the west of Ireland. Closer to the Tees, it's not too hard to find these orchids on the limestone pavements around Ingleborough in Yorkshire. But at Bishop Middleham, dark-red helleborines grow in drifts and lush clumps across the whole quarry. It's thought that 90 per cent of the British population grows in this one tiny, but very special, spot. It's almost impossible to decide which clump to photograph – any one of them would win a prize in an orchid show – not that the flowers are in-your-face spectacular. They're only half an inch across and an inconspicuous dark red in comparison to the fluorescent pinks and magentas of the other orchid species that share the short turf. Through a macro lens, however, they have their own elegance, their dark-red petals and sepals offset by bright-yellow anther caps at the centre of each flower.

On a hot day in late June, the quarry is alive with butterflies and day-flying moths. Burnet companions, burnet moths and Mother Shipton moths compete with common blues and grizzled skippers for nectar or busy themselves laying eggs on their favoured foodplants. However, I've timed my visit here in the hopes of meeting a much rarer butterfly – one that I haven't seen before – the northern brown argus.

As its name suggests, this is a northern species which doesn't live much further south than the colony at Bishop Middleham quarry. In the southern half of Britain, it's replaced by the brown argus, a much more common though very similar-looking butterfly, with almost identical brown upperwings, framed with a line of bright-chestnut spots. The brown argus flies in some abundance on the warm grasslands of the Cotswolds and on the chalk downs, as well as many other flower-rich grassy sites. In the past its range didn't extend further north than a few colonies in Yorkshire, a little way to the south of the Tees, but this little butterfly is currently spreading north at a fast pace. In most of Britain only one of these species is likely to occur, which simplifies the task of distinguishing between two almost-indistinguishable butterflies. In Scotland, northern brown arguses also sport a tiny white dot in the centre of the upper side of each forewing – a handy confirmation that you've found one of these special little butterflies. However, a slightly different variant flies in northern England, and this one lacks the distinguishing white spot.

This is, to say the least, very unhelpful on the part of nature. In North-East England, the Tees Valley may once have been the rough boundary between these two species of butterflies, but since it's almost impossible to tell them apart, it's hard to judge how much overlap there is between the two species and how much a warming climate has allowed the brown argus to push into northern brown argus territory. Indeed, those colonies in Yorkshire now identified as brown arguses were once assumed to be the northern species. Lower down the Tees, in 2009, one single brown argus turned up on an industrial site monitored by INCA and, not much more than a decade later, there were strong colonies of the southern species on several sites around the estuary, and on both sides of the river.

Genetic studies (the only certain way to know which is which) indicate that both species are moving north, the northern species retreating ahead of the advancing southern one. But where they overlap, the two species also hybridise, creating a zone of even greater confusion. For example, northern brown arguses from North Wales are now thought to be brown arguses that nevertheless, through past hybridisation, carry some northern brown argus genes. It's all mind-achingly confusing. I soon find a few brown-and-chestnut butterflies fluttering around rockroses on the quarry floor and on the basis that Bishop Middleham is a well-known site for northern brown argus, I'm going to label my photographs as this species and give it a mental tick. Job done.

Walking back to the entrance, I pause by the low cliff face which is now busy with the comings and goings of sand martins, twittering and churring excitedly as they ferry beakfuls of insects back to hungry chicks. They've excavated dozens of tightly packed, deep tunnels in the loose, sandy rock and the whole cliff face is now alive with diligent parents toing and froing. Back in 2002, though, it was another hole-nesting bird that caused a real stir here, when a pair of multi-hued European bee-eaters decided to set up home in the quarry.

Further south in Europe, bee-eaters rear their chicks in colonies as busy as those of sand martins. I've seen colonies in Hungary where several hundred birds had excavated their tunnels along an old riverbank and flew, like sleek fighter jets, back and forth to nearby meadows to catch bees. On returning, each bird often perched on a branch, where it beat the poor insect senseless to remove the dangerous sting, before disappearing into the nest tunnel to feed its chicks. Stray pairs of these colourful birds have occasionally arrived in Britain over the last century and have often attempted to breed – with varying success. The birds at Bishop Middleham, however, succeeded in fledging a couple of chicks.

In more recent years, larger groups have been arriving on our shores as climate change turns Britain into a more hospitable place for these Southern European migrants. In 2017, seven pairs nested in Nottinghamshire, and in 2022, another seven pairs bred in an old quarry in Norfolk. Several birds returned to the same site in 2023, the first time in Britain that they've returned to a previously occupied site – perhaps the vanguard of a permanent breeding population. In all these cases, the rainbow visitors drew many thousands of admiring birdwatchers. At Bishop Middleham, the Durham Wildlife Trust set up a carefully monitored viewing station at a discreet distance from the nest, which ended up hosting 15,000 visitors over the season. It's hard to imagine that this quiet backwater near the source of the Skerne could ever be so crowded. As I make my way back to the car, I haven't seen a single soul all day.

Retracing the Skerne back towards the Tees takes us through the town of Darlington, which straddles the River Skerne, and into yet more botanical adventures, this time closely tied to the story of the Stockton and Darlington Railway, in whose tracks we've been more or less walking on our journey so far. In the nineteenth century, Darlington was home to several important Quaker families who helped shape the last few centuries of history along the Tees. We met Joseph Pease earlier in our journey, the man who originally established Port Darlington at the end of an extension to the Stockton and Darlington Railway, a settlement which soon became Middlesbrough. But the world's first public railway could never have happened without some serious financial backing for the whole groundbreaking venture, much of which came from another of Darlington's Quaker families – the Backhouses.

The Backhouse family, like many others of this faith, found their way into banking thanks to bigotry and prejudice. The ideas that gave rise to the modern Religious Society of Friends, or Quakers, arose in the aftermath of the English Civil War, when, after years of hardship and bloodshed, there seemed little point choosing between the old overbearing monarchy or the new – and equally overbearing – Lord Protector Oliver Cromwell, both of whom insisted on their own closely prescribed ways of worship. Faced with an impossible choice, Quakerism took the sensible option of abandoning both the clerical hierarchy and an ordained form of church service in their entirety. Following the Act of Tolerance in 1689, Quakers were granted the freedom to practise their own brand of worship but, as Non-Conformists, they were still barred from many top-level professions such as those in law, in public office or in the universities.

That left business and banking – and Quaker banks turned out to be some of the most successful in the country. Quakers valued honesty, diligence and sobriety, and certainly never partied the nights away, so Quaker-owned banks, not surprisingly, soon became very popular. In Darlington, James (I) Backhouse (1721–98), together with his two sons, Jonathon (1747–1826) and James (II) (1757–1804), founded Backhouse's Bank in 1774. Eventually, this bank amalgamated with two other Quaker banks based in London to form Barclay's Bank. The original Backhouse's Bank in Darlington is therefore now Barclay's Bank, but it still stands where it always has, in a prominent spot on High Row in the centre of town. However, beyond their interests in banking and business, the Backhouses also had an abiding interest in botany and gardening.

With a flair for the predictable, a descendant of James (II) in each generation was also named James. James (III) (1794–1869) and James (IV) (1825–90) were major players in the discovery of the unique flora of Upper Teesdale, so we'll meet them again further up the valley. Jonathon's son William (I) (1779–1844) was also a botanist, while his son William (II) (1807–69) bred daffodils to supply a Victorian and Edwardian craze for these plants, a venture that proved almost as lucrative as banking.

One way to create attractive new varieties of plants is to force them into creating seeds, which contain multiple copies of the plant's basic set of chromosomes – so-called polyploids. William Backhouse was the first to create a triploid daffodil cultivar (with three times the basic chromosome number), which he called 'Empress'. He also created the first tetraploid cultivar (with four times the basic number of chromosomes), called 'Weardale Perfection'. In all, the Backhouses bred 950 cultivars, although fewer than 20 are still available to gardeners today, and they're really hard to find. However, both the 'Empress' and 'Weardale Perfection' cultivars are still (if very rarely) grown.

In the 1860s, on the other hand, *Narcissus* 'Empress' was one of the most sought-after daffodils because polyploids are generally larger and more vigorous, making them highly desirable as garden plants. *Narcissus* 'Weardale Perfection' likewise proved so popular that in the 1900s, real enthusiasts were prepared to pay £12 (£1,400 at today's values) for a single bulb. William worked full-time as a banker so had to rise early enough each day to carry out his breeding experiments before catching a train to work. Luckily, his family had helped set in motion the very rail network that allowed him to indulge his passion and still get to the family bank on time to earn his keep.

Early in the nineteenth century, several other Quaker families had already invested heavily in the port of Stockton by straightening the Tees below Stockton to speed the journey from the port to the North Sea. Now they needed to increase trade through the port to get a return on their investment. One of the main exports was coal from the coalfields in the southern part of County Durham, so this was the obvious trade to target. Several schemes were discussed, including a system of canals connecting the collieries to the river at Stockton and a rail system, originally imagined as horse-powered. Similar horse-drawn lines already existed elsewhere, but Edward Pease, father of Joseph, made the visionary suggestion of using steam locomotives on the line instead. Furthermore, the line was conceived as a public railway, for which anyone could buy transport, either for goods or as a passenger. And so, in 1818, the Stockton and Darlington Railway Company was established.

Pease employed a Tyneside engineer, George Stephenson, to design a suitable locomotive. Stephenson also suggested an easier alternative route for the line which bypassed Darlington.

However, since most of the leading lights promoting the project, as well as those bankrolling it, were based in Darlington, Edward Pease informed Stephenson in firm but polite tones befitting a Quaker: 'George, thou must think of Darlington; remember it was Darlington that sent for thee.' So, despite it being a less-direct route from the coalfields around Shildon to Stockton, it ended up as the Stockton and Darlington Railway, the world's first public railway.

The inaugural journey took place on 27 September 1825. So many people turned up to ride the train from Shildon to Stockton that it was badly overcrowded. Along the way, one wagon lost a wheel, and there was a further hold-up when the locomotive needed running repairs. Even on day one, those cherished traditions of the British railway system – of overcrowding and long delays to services – had been well established. Yet, despite travelling at only a little above walking pace, the first journey was deemed a success. The vision of those men from Stockton, Yarm and Darlington would kick-start a transport revolution not just in Britain but around the world. The modern age of speedy mass transport began, albeit at a stately pace, on the banks of the Tees.

Railways were not the first transport revolution to transform Britain. Leaving Darlington, the river and the Teesdale Way pass under the busy A1M – the Great North Road. The path follows the loops of the last of the big meanders on the river and then arrives at a much older Great North Road, now marked by little more than an untidy pile of stones in the middle of a field. These stones, at Piercebridge, were once the foundations of a bridge that carried a Roman road called Dere Street across the Tees, at a time when Britain first entered the orbit of a European economic community, which inevitably brought many changes to these islands.

In 55 BC, Julius Caesar made two forays into southern Britain from the conquered territory of Gaul. Although he didn't establish a permanent Roman presence here, in the century that followed, trading links between Britons and the Romans became stronger and some British tribes, such as the Atrebates, grew ever more pro-Roman. Other neighbouring tribes, however – the Iron Age equivalent of what used to be called the Brexit Party – were less enthusiastic about such close ties with a European superpower. It could only lead to trouble. Beginning in AD 43, and partly to consolidate his position in Rome, Emperor Claudius decided to extend the Roman Empire to the British Isles. Key to the success of this invasion and to the subsequent control of many – often hostile – tribes was efficient transport and communication along a system of fast roads – Britain's first motorway network. Dere Street ran from the northern legionary headquarters at Eboracum (York), which centuries later would become the Viking regional capital of Jorvik, north across the Tees, roughly parallel to Cade's Road, which we crossed further downstream at Middleton One Row. Dere Street then ran on through what is now Northumberland, past Hadrian's Wall as far as the Antonine Wall, which crossed Scotland between the Firth of Forth and Firth of Clyde.

A wooden bridge was built across the Tees around AD 80–90, near present-day Piercebridge, but this seems to have been washed away, probably in one of the river's forceful floods. A second bridge, with more robust stone foundations, was constructed in AD 150 and these foundations are still visible today. However, they now lie in a field about 100 yards to the south of the river, which has changed course dramatically since Roman times. Nearby are the remains of a Roman civilian settlement, probably built around AD 100, and a later fort, indicating that the Romans soon brought their lifestyle to the Tees Valley.

Walking around the jumble of stones of the Roman bridge at Piercebridge, it feels like we've reached a turning point in our journey. We're leaving the more recent history of the Industrial Revolution that has shaped the lower river and we're about to walk through more rural landscapes which have a far longer history. This change in the character of the river at Piercebridge was also evident to another traveller along the Tees in the late nineteenth century.

In 1890, one Aaron Watson (1850–1926), a journalist, made the same journey as me but in the opposite direction. Heading down from the high fells and the upper Tees, he considered that the river's beauty began to diminish once he passed downstream of Piercebridge, and his descriptions become increasingly disparaging the closer he got to the river-mouth.

> By the time it reaches Stockton Bridge the Tees has been transformed from one of the most wild and lovely to one of the most tame and repellent of existing rivers. Its soiled waters henceforth flow between banks of blast-furnace slag; unpleasant odours float about its shores.

> It is an unlovely Tees that the eye alights upon since the smoke of the blast furnaces came into sight. It would scarcely be possible for a river so beautiful in its upper reaches to undergo a more surprising and spirit-depressing change.[*]

By knowing where to look, we've managed to find enough natural beauty to lift our spirits even amid the industrial landscapes of the lower river, although I know what Watson means. However, travelling in the opposite direction to Watson means that we've now left the large conurbations that flank the lower river and estuary and instead we're surrounded by agricultural land – land that has supported the many different cultures that have lived in the Tees Valley over the millennia and who have each played their role in shaping both landscape and nature.

It's easy to see the landscape around Piercebridge, of verdant pastures or golden fields of wheat and barley, as both timeless

---

[*] Aaron Watson. *The Rivers of Great Britain: Rivers of the East Coast*. Cassell and Company Ltd. 1889.

and a welcome relief from the urban and industrial sprawl that hems in so much of the lower Tees. A chance to breathe fresh air and immerse yourself in nature. However, this 'green and pleasant land' is as much the product of industrial processes, in the form of intensive farming, as the landscapes surrounding the Tees Estuary, and in many cases is even more inimical to nature. The sharp contrast that Watson drew between the character of the river above and below Piercebridge was, even by the late nineteenth century, superficial, and it is even more so today. Many people still find it hard to believe that our rural landscapes are some of the most nature-poor of any nation and instead celebrate this classic image of the British countryside, a view which has a long tradition, with its roots in the late eighteenth century, among the Romantic poets and artists, several of whom spent considerable time in the valley of the Tees.

In his passionate evocation of the English countryside, William Wordsworth (1770–1850), who spent time on the Tees at Sockburn, is often regarded as one of our earliest modern environmental thinkers, although when he was writing, the countryside that he exalted was already being as rapidly stripped of natural worth as the blasted industrial landscapes. At the same time, artists like J. M. W. Turner (1775–1851) portrayed similarly romantic views of the English countryside, as well as darker scenes of how the world was being changed by the Industrial Revolution. Turner also visited the Tees Valley, first in 1797, and by 1831 had made three further visits. While he was painting wild scenes along the river upstream from Piercebridge, crowds in Darlington were marvelling at the start of the railway age and, further downstream, blast furnaces were being constructed along the lower Tees.

Turner's painting *The Junction of the Greta and Tees*, which today hangs in the Ashmolean Museum in Oxford, portrays the Meeting of the Waters, where the River Greta flows into the Tees, some eight miles upstream from Piercebridge. It is a somewhat idealised view of this quiet spot, but the mood of the place is much the same as that captured in paint by Turner two centuries ago. Turner was also commissioned to provide illustrations for a long poem by the novelist and poet Walter Scott (1771–1832), another famous visitor to Teesdale. While staying at Rokeby Hall, a Palladian country house a little way upstream on the River Greta, Scott wrote *Rokeby*, an epic poem set in a picturesquely evoked Teesdale during the turmoil of the English Civil War. The gorge near Rokeby Hall, through which the River Greta runs before joining the Tees, is still much as Scott described it.

> The river runs with very great rapidity over a bed of solid rock, broken by many shelving descents, down which the stream dashes with great noise and impetuosity ... The banks partake of the same wild and romantic character, being chiefly lofty cliffs of limestone rock, whose grey colour contrasts admirably with the various trees and shrubs which find root among their crevices.

Rokeby was one of Scott's most popular poems, and between them, Scott and Turner put Lower Teesdale firmly on the tourist map. As industry spread along the lower reaches of the Tees, tourists flocked to Teesdale for spiritual refreshment. The times were changing, and this was driven in no small part by the works of the Romantics.

It is to another artist of the Romantic era, William Blake (1757–1827) – a poet and painter with, to say the least, a very idiosyncratic view of the world – that we owe the phrase 'green and pleasant land'. It appears in his poem 'And Did Those Feet in Ancient Time' – now more famous as the hymn 'Jerusalem', quoted at the start of this chapter – where he describes a supposed visit to England by Jesus Christ, drawing a stark comparison between England's idyllic rural landscapes and the 'dark satanic mills' of the burgeoning Industrial Revolution.

We've already seen how the rapid spread of industry in the

nineteenth and twentieth centuries swept nature aside or buried it under belching smokestacks and factories where men and women laboured unceasingly – and for a pittance. And yet nature was in retreat across almost all the rural landscapes of Britain long before this time. Indeed, had Jesus really walked on our pastures green two thousand years ago, he would have seen a world already drastically reshaped by many generations of Iron Age farmers. However, since Blake's time, our countryside has changed at an ever-faster pace, and the retreat of nature has only accelerated. Today, our farmlands may look green – but pleasant they certainly are not.

Wandering back from the scattered stones of the long-vanished Roman bridge to the modern bridge that crosses the Tees at Piercebridge, I find myself reflecting on the long history of the English landscape, and on how our countryside has been changed by waves of new people bringing different ways of managing the land along with new plants and animals. We've already looked back as far as the Anglo-Saxon kingdoms bordering the river and to the subsequent changes wrought by Viking newcomers, but the layers of history go far deeper. Even when the Romans were busy building Dere Street across the Tees Valley, the landscapes through which the road passed had already been altered by hundreds of generations of people, and by vast natural forces. Leaning on the parapet of the bridge at Piercebridge with a view up and down the river and just a little less than halfway through our journey, this seems like a good place to pause and to place the remainder of our journey in a wider context, that of the history of rural Britain since the end of the last Ice Age.

Interlude

After the Ice

A round 20,000 years ago, the last glacial period reached its greatest extent. Ice covered all of the northern half of Britain and where I'm standing today was at the edge of a vast ice sheet which, in places, was more than half a mile thick. The ice often blocked the Tees itself, and when it did, a vast ice-dammed lake, called Lake Tees, flooded the whole of the present lower Tees Valley, from Yarm to the North Sea. An ice age consists of a series of these cold glacial periods, each around 100,000 years in duration, interspersed with warmer 'interglacials', lasting perhaps 10,000 years or so, during which the ice retreats and warmth and life return to the land. A proto-Tees probably flowed in these mild intervals, its banks grazed by animals such as hippopotamuses that we wouldn't usually associate with the Tees Valley. Straight-tusked elephants, twice the size of modern species, along with lions and hyenas, no doubt also roamed the valley. But when the ice returned, the slate was wiped clean.

Around 14,000 years ago, the climate warmed once more, with startling speed. Water flowed again in the Tees Valley, but then, after just a thousand years or so, the cold returned. Glaciers renewed their advance from ice fields capping the high Pennines, and cold tundra clothed any land that was free of ice. Fortunately, this turned out to be merely a blip in the warming trend, probably caused by a temporary shutdown in the

219

Gulf Stream, part of an ocean-wide pattern of circulation in the Atlantic. This circulation is driven by a 'pump' located in a small area between Iceland and Greenland. Here, cold, dense, salty water descends to the ocean depths. The sinking water draws a current of warmer water, the Gulf Stream, north from the Caribbean which bathes our shores and makes Britain far milder than its northerly position would otherwise allow.

When the climate warmed at the end of the last glacial period, vast quantities of freshwater from the melting ice sheets flowed into the North Atlantic, diluting saltwater and shutting off the pump that powered the circulation around the whole North Atlantic basin. Our central heating failed – as it may well do again as a warming climate caused by our overuse of fossil fuels melts the great ice sheet that still cloaks much of Greenland. It should give us all pause for thought that Britain lies at roughly the same latitude as the bleak Canadian province of Labrador, on the other side of the Atlantic.

Nevertheless, the cold blip (called the Younger Dryas Period by geologists) was short-lived, and between 11,000 and 12,000 years ago, warmth began to return. This period saw another strange collection of animals occupy Britain. Herds of reindeer, wild horses and saiga antelopes roamed the tundra, while steppe pikas (small rabbit-like creatures), lemmings and ground squirrels grazed the grasses and sedges. It was a unique era in environmental history; in our own time, nowhere on Earth does such a motley collection of creatures still live side by side.

As the climate continued to warm, animals like these, which still thrived in Britain at the very beginning of the current warm period, abandoned our shores, although a few did manage to hang on. A tiny snail called *Vertigo genesii* is today confined to the Arctic or to high mountains in Europe. It was, however, once common across much of Britain in the late glacial period and, until recently, it was assumed to have followed the reindeer and lemmings to more northerly climes – until a single population of this snail was discovered living very happily in Upper

Teesdale. This minute snail is hardly a showstopper, but all the same, it's a living link with a very different Britain, as the land was newly revealed from beneath the ice – a glimpse of the last time Britain was truly wild.

Once a warmer climate finally took hold, most of the Arctic creatures disappeared from our shores. However, at this point in our history, we were still joined to mainland Europe by a broad land bridge across what is now the shallow Dogger Bank, beneath the southern North Sea. So, new animals and plants tracking the warming climate northwards from refuges in Southern Europe could simply walk into Britain to replace the tundra species that were themselves migrating north. That remained the case until about 8,000 years ago, when rising sea levels, driven by melting ice and sinking land,* finally flooded the land bridge and turned Britain into an island. Anything that hadn't already made it here, or couldn't fly, was excluded, at least until people began crossing the Channel more frequently in boats. Quite a few animals didn't arrive in Northern Europe until after the land bridge to Britain was flooded, so, in this way, we missed out on sharing our islands with two species of white-toothed shrews, garden dormice, pine voles and characterful stone martens, renowned in Germany for exploring under the bonnets of parked cars, where they often chew through a few vital cables.

Neither our flora nor our fauna was as rich as those of mainland Europe, although by the time Britain was cast adrift, we did at least have a good complement of trees. As the climate warmed, hardy trees and shrubs, such as birch, aspen, alder and hazel, replaced tundra with open woodland. Next, pine woods began to develop, followed by woodlands composed of today's familiar broadleaved trees – oak and ash, lime and

---

* With the weight of the ice removed from northern Britain, the land began to rise (a process known to geologists as isostatic uplift). Consequently, like a giant seesaw, the land in the southern half of Britain began to sink.

hornbeam. As the climate warmed further, the more cold-hardy trees retreated northwards to be replaced by warmth-loving species from the south. Over millennia, waves of different types of forest marched across Britain. These extensive primeval forests are known by the evocative name of 'wildwood', which, for those of us reared on the tales of the Brothers Grimm, conjures up a dark, endless forest full of terrors – an easy place to get hopelessly lost. However, no one is really sure what this wildwood looked like.

In 2000, the Dutch ecologist Franciscus Wilhelmus Maria Vera published a book that revolutionised the concept of the wildwood and generated a healthy debate that continues to this day.* His view is that the wildwood was nothing like the vast tangled forest of European fairy tales, but was instead an ever-changing mosaic of closed forest, scrub and open grassland, created and maintained by an abundance of big grazing animals. Derek Yalden, of the University of Manchester, tried to put some numbers on this megafauna, using the current populations of large mammals in places like Poland's ancient Białowieża Forest as a guide. According to these calculations, Britain could have been home to 100,000 aurochs and a million wild boar, along with more familiar red and roe deer. The grazers and browsers were hunted by 20,000 wolves and 10,000 Eurasian lynx.

However, because we became an island not long after the end of the glacial period, Britain's big grazing fauna wasn't as diverse as that found in Europe, where Vera carried out his research. Although European bison had lived in Britain during previous interglacials, they didn't return after the last retreat of the ice. In addition, wild horses became extinct in Britain early in the current interglacial. So, the story of Britain's wildwood might be a little different from that of Europe's. However, the

---

* Franciscus Wilhelmus Maria Vera, ed. *Grazing Ecology and Forest History*. CABI publishing. 2000.

remains of beetles from deposits ranging in age from 9,000 to 6,500 years ago suggest that substantial tracts of open ground existed during this period.

In any case, the migration of animals to our shores in the wake of the ice was accompanied by waves of Mesolithic hunter-gatherers. Collections of beetle remains from deposits dating from less than 6,500 years ago show that the insect fauna was changing as species more frequently found in closed forests increased in abundance. Big grazers were tempting targets, and it's been suggested that generations of Mesolithic hunters had begun to reduce populations of mega-herbivores to such an extent that forests began to reclaim some of the grasslands.

The biggest of the grazers in Britain was the aurochs, the wild ancestor of domestic cattle. The bulls in particular were impressive beasts, standing close to six feet in height. Unlike domestic cow breeds, aurochs had long legs and an athletic build and were agile and swift, which must have made them challenging to hunt. Yet hunted they were. In the North-East, their remains are abundant at Starr Carr, just to the south of the North York Moors, a Mesolithic site dating from 9,000 years ago, not long after the ice had retreated. At the time, Starr Carr sat on the shores of the huge Lake Flixton, which filled the modern Vale of Pickering after ice blocked its drainage to the sea. Venison must also have featured on the menu for the people of Starr Carr because elk, the world's largest deer, love marshy terrain like this and they existed in large populations around the lakeside settlement.

These big grazers didn't survive for long. In Britain, aurochs probably became extinct sometime in the Bronze Age, perhaps between 1500 and 1000 BC, although they survived in Europe for much longer. In France, they were hunted by royalty until the ninth century, and the very last of their kind died as late as 1627 in the Jaktorów Forest, not far from Warsaw in central Poland. Despite once ranging from Britain to the Far East, aurochs are now globally extinct. Elk, too, may have vanished from Britain in the Bronze Age, although some could have survived into the Iron Age or even later in the more remote parts of Scotland. However, unlike aurochs, elk still survive in parts of Europe as well as across North America, where they're known as moose.

The Mesolithic hunter-gatherer communities certainly had some impact on Britain's landscapes, but changes accelerated in the Neolithic period, beginning a little over 6,000 years ago, when farming came to the British Isles. This Neolithic Revolution arrived with waves of new colonists from Europe, who brought with them a culture based on domestic animals and plants that had originated in the Near East several thousand years earlier. Farmers need fields, both to graze livestock and to grow crops and hay, so forest clearances escalated. The remains of beetles from this period tell us that open-ground species once again thrived, including dung beetles, which had a plentiful food supply from the ever-increasing numbers of livestock. Even if grazing hadn't been responsible for creating and maintaining large tracts of grasslands in the original wildwood, it certainly was now.

From the Neolithic to the present day, the story of the British countryside has been one of ever-more-intensive farming and an ever-less-natural landscape – with just a few short-lived reversals. By 2000 BC, at the start of the Bronze Age, large parts of Britain were covered in fields, creating a landscape that wouldn't seem too out of place to modern day-trippers to some parts of our countryside. By 1000 BC, at the start of the Iron Age, wheeled vehicles were common, and the mechanisation of farming was well underway. The North of England was prime real estate and home to the most extensive Iron Age tribe in England – the Brigantes. Their vast territory was centred on Yorkshire but also stretched over much of modern County Durham, Northumberland and Lancashire.

Changes to the landscape and the slow retreat of nature continued through the Iron Age, although it's thought that Iron Age people may still have regarded wild nature with some form of reverence. They lived largely off the fruits of their farming and there's little evidence to suggest that wild game or fish were hunted for subsistence on a regular basis. Where remains of wild creatures have been found at Iron Age sites, they're often in a ritual or religious context. But even this attitude began to change after AD 43.

When the Roman legions of Claudius began building their roads north through the valley of the Tees, much of Britain's forest had already been cleared, and they would have crossed large tracts of rich farmland. Although the arrival of the Romans may have had little direct effect on the landscape, they brought with them new ideas, along with new plants and animals. In the immortal words of John Cleese (as Reg in *Monty Python's Life of Brian*): 'What have the Romans ever done for us?' Well... they introduced around fifty new food crops to British farms, including onions, parsnips, cabbages, carrots, leeks and asparagus, along with domestic apples – a welcome respite from the bitter native crab apples – and walnuts, pears and grapes.

Other fruit and veg found at Roman sites include lentils, figs and mulberries; the latter may have been imported at first, but eventually mulberry trees were widely grown in Britain. The legions were apparently also fond of cherries. Wild cherries are native trees of British woodland, but domestic cherry trees originated in the far south-east of Europe and had only arrived in Rome itself shortly before the Empire encompassed Britain. Even so, it's said, probably without much basis in fact, that you can trace the routes of long-vanished Roman roads by the lines of cherry trees that sprang up from the seeds spat out by marching soldiers.

The Romans also brought new animals, although it's surprisingly hard to say exactly when many of our introduced animals became established in Britain. The Romans have been credited with bringing brown hares, pheasants and fallow deer to our shores, and possibly even rabbits. However, there's some evidence to suggest that brown hares arrived here during the Iron Age.* More usually, rabbits and fallow deer are assumed to have arrived with the Normans, a full millennium after the Romans, but a recent find at a Romano-British villa in Fishbourne in Sussex suggests that the Romans imported at least a few of both these creatures. Fishbourne Palace was a luxurious dwelling, the largest such palace in the whole of the Roman Empire to the north of the Alps, and here both the rabbits and deer were precious exotics, highly prized as status symbols.

The Romans kept all kinds of wild animals in captivity. Rabbits and hares were confined in large enclosures called *leporaria*, while deer were kept in extensive, securely fenced parks. Although hard to judge at the remove of two millennia, it seems that nature was losing its sacred status as Roman ideas spread, or, at the very least, the Romans preferred a tame and controlled natural world – as well as an equally tame and controlled world among the subjugated people of the Empire, although the latter wasn't always so easy to achieve.

In the Tees Valley, just a few miles from where I'm standing on the bridge at Piercebridge, lies Stanwick, one of the largest fortified and enclosed Iron Age settlements in Europe. These forts were known to the Romans (and to modern archaeologists) as *oppida*. Dere Street passed close to the fort at Stanwick, where excavations have uncovered a wealth of Roman luxury goods such as wine amphorae and rare glass and tableware objects. It seems that the locals (or at least some of them) had thrown themselves wholeheartedly into trade with the newcomers, enabling them to put on shows of opulence that enhanced their status. It's possible that Stanwick was the capital of Cartimandua, queen of the Brigantes, who was a staunchly

---

* The mountain hare, *Lepus timidus*, is native to Britain and although now confined to Scotland and the Peak District, was once more widespread. It's very hard to tell the difference between the remains of this species and the introduced brown hare, *L. europaeus*, in archaeological deposits.

pro-Roman Briton. Unfortunately, her husband Venutius was far less enamoured with the invaders and his machinations led eventually to attacks by the Brigantes on the local Romans, and to civil war within the tribe. With Roman assistance, Cartimandua eventually defeated her (by now) ex-husband, and the Roman rural lifestyle became established in the Tees Valley, not far from the ever-watchful legions at Eboracum (York).

Roman rule remained for a little over three centuries, until AD 383, when the legions withdrew from the North of Britain to combat internal struggles within the Empire. The Tees Valley was once more left under the control of local chieftains. Just a few decades later, as the Visigoths sacked the very heart of the Empire – the city of Rome itself – all the Roman legions were recalled from Britain to defend what was left of the Western Roman Empire.

During the fifth century, a period of our history for which the written record is extremely sparse, much of Britain was settled by Angles and Saxons, Germanic invaders from an area that is now southern Denmark and northern Germany. And these people brought a new era to the farmlands of Britain.

Around Saxon settlements, the land was farmed intensively using a three-field system. These fields were often extremely large and open, reminiscent of a modern, intensively farmed landscape, but planted sequentially with different crops in a sophisticated crop-rotation system. First, a field was sown with cereals such as wheat or rye, then the following year with beans, peas or lentils. Finally, in the third year, the field was left fallow, before the cycle began again. The three fields were planted out of step so that surrounding each village was a vast field of cereals, one of pulses and one lying fallow. Cereals deplete the soil of nutrients, but legume crops such as beans or peas, with the help of symbiotic bacteria in their roots, turn nitrogen from the air into nitrates which enrich the soil. So, rotating crops like this replenishes and maintains the fertility of the fields.

Some time ago, I visited a working example of these Saxon open fields which had survived at Laxton in Nottinghamshire, a heritage proudly announced on the village signs – 'England's last open field village'. It's a vast treeless landscape from which nature seems to have been largely banished, although winter stubble and the fallow field does provide food and cover for some birds. Even so, a survey in 1970 revealed that there were far fewer birds on these open fields than on neighbouring farms. With the arrival of the Anglo-Saxons, further depletion of nature on our farmlands had begun – although even worse was yet to come. Those birds thriving on Laxton's open fields, mainly skylarks, lapwings and grey partridges, are all now severely threatened across most of the rest of our countryside. The population of skylarks, for example, has fallen by nearly 60 per cent in just four decades, between 1970 and 2010. Such open fields persisted and spread throughout medieval times, beginning a long period when nature found it much harder to survive in farmland.

In the middle of the eleventh century, the Normans invaded and brought more changes to both the human and natural history of Britain. We've already met several Norman nobles who were granted lands along the Tees as Norman kings tried to exercise control over northern England. Norman kings also declared vast tracts of England to be Royal Forest, where game such as deer and wild boar was strictly protected and where hunting was reserved for the king or favoured noblemen. To bolster game populations, the Normans also imported large numbers of fallow deer from Europe.

Forests and forest law arrived in Britain with William the Conqueror and soon proliferated. By the time of the Domesday survey in 1086, there were already twenty-five forests, and afforestation continued into the twelfth century under King Henry I. He declared Epping and Sherwood Forests and seems to have been responsible for the introduction of large numbers of fallow deer to Britain. The few captive fallow deer at major

Romano-British palaces almost certainly died out after the Romans left, so our current population is likely to have originated from Norman introductions. A great many deer were also kept in enclosed deer parks attached to large estates, for which they provided ample meat. There's a fine example of such a deer park in the Tees Valley, not far from Piercebridge.

On the banks of Langley Beck, three miles upstream from its confluence with the Tees, the magnificent Raby Castle dominates the landscape. It's surrounded by a large deer park with herds of both fallow and red deer. Red deer are native to Britain, whereas fallow deer originated in Anatolia, from where they were transported across Europe and eventually into Britain to fill Norman deer parks. Raby Castle was built in the fourteenth century, by which time deer parks had been growing ever more popular for three centuries. When Raby Deer Park was established 600 years ago, it was one of around 3,000 similar parks across the land, housing substantial herds of fallow deer.

Although not all forests belonged to the king, most did – and they became a way for the king to control the local nobles. As such, afforestation was curtailed after Magna Carta in 1216, as part of the restrictions placed on the king's power by the barons. By this time, however, there were already at least 143 Royal Forests in England. Deer and other game species were integral to medieval life and politics, so anything – other than the king and his guests and allies – that ate deer was not tolerated. The big predators were in retreat. Brown bears once roamed widely through British woodlands and may have still survived in small numbers in Scotland when the Romans arrived, but they vanished soon after that. It's often assumed that lynx were likewise extirpated early in our history, although recent finds suggest that these deer-hunting cats actually managed to survive, at least in low numbers, in a few scattered locations until the sixteenth century.

Wolves, smart and adaptable, survived in large numbers for a lot longer, but after the Norman Conquest and the arrival of a culture that sought to exercise an even greater control over nature than the Romans, all-out war was declared on these predators. Nobles were given land on the understanding that they would keep it free of wolves. King Edward I, known as the Hammer of the Scots, is less widely known as the Hammer of the Wolves. In 1261, he engaged a professional wolf-hunter to rid the counties of Gloucestershire, Herefordshire, Worcestershire, Staffordshire and Shropshire of wolves, some of the few remaining strongholds of wolves in the southern half of England. Many such *luparii* were employed across Britain during the twelfth and thirteenth centuries to exterminate wolves, although Edward I was particularly diligent in this task, to the extent that wolves had become scarce by the end of his reign.

By the end of the thirteenth century, wolves were rare in southern England, although they still roamed in the North. The last reliable record of wolf-trapping in England comes from Whitby Abbey on the Yorkshire coast, where, sometime between 1394 and 1396, monks paid 10s 9d for fourteen wolf skins. Perhaps a few still roamed the wilder parts of Yorkshire in the fourteenth century. Wolves hung on in out-of-the-way places elsewhere in Britain until the sixteenth century, by which time they were so rare that they acquired the status of an exotic animal. In 1599, one was housed in the animal collection at the Tower of London, where it was described as 'the only one in England'.

Over-exploitation of game animals resulted in dramatic falls in their numbers too. For his Christmas dinner in 1251, King Henry III ordered 200 wild boars from the Forest of Dean and another 100 from the Forest of Pickering in Yorkshire. In 1260, only twelve wild boars were sourced from Dean, and after this, records cease. Even smaller creatures, such as martens, stoats and squirrels, were targeted by a growing market economy. There are records from 1344–45 of the belly fur of nearly 80,000 red squirrels being used to trim clothes. When

King Richard II's widow, Isabella of Valois, returned in mourning to France, her sombre garments still managed to incorporate nearly 46,000 squirrel skins. In a market economy, prices rise when a commodity becomes scarce, which, in the case of animals hunted in the wild, makes it financially worthwhile to track down even the last few individuals.

We Britons have been particularly effective at eliminating nature in part because we live on an island. Elsewhere in Europe, animals ranged widely over the whole continent and survived in remote wildernesses from where they could later recolonise areas from which they'd been wiped out. In Britain, there were fewer places to hide, so it proved much easier and a lot quicker to eliminate most of the larger species entirely. Even by the Middle Ages, Britain was well on its way to becoming the nature-poor island that we live on today.

The free-market economy sees wild nature as an infinitely renewable source of commodities and so, over the centuries, has inflicted massive damage on the natural world. However, the consequences of changing farming practices have been just as severe. In the medieval period, sheep wool was one of our most important exports, so much so that the Speaker in the House of Lords sat on a woolsack,* to remind the assembled nobility of the economic importance of these animals. Yet the impact of sheep on Britain's economy was as nothing compared to their impact on our ecology. In the late nineteenth century, the Scottish-born American ecological thinker John Muir (1838–1914) famously described the sheep that grazed in California's Sierra Nevada Mountains as 'hooved locusts', but even 600 years before that, by 1300, there were already 20 million of them overgrazing grasslands and pastures across Britain. For comparison, there are 23 million sheep in Britain

---

* You can't believe anything in politics. It turned out that it was actually stuffed with horsehair, although it was restuffed in 1938 with wool sent from across the Commonwealth.

today, so the serious damage to the landscape caused by too many nibbling and trampling sheep must have been prevalent even by the Middle Ages. One reason they're so destructive is that sheep aren't native to Europe. Domestic sheep originated from wild mouflon in South-West Asia, so sheep were never part of our evolving ecology after the Ice Age. Cattle and pigs at least had wild ecological antecedents in Britain.

In the late seventeenth and early eighteenth centuries, as the human population rose, so new intensive ways to increase arable production were developed. Jethro Tull (no, not the band) came up with a seed drill that was more efficient in sowing seeds than the old method of scattering handfuls of seeds across the field. He also developed a horse-drawn hoe that made the back-breaking task of manually removing weeds a lot easier. At the same time, Charles Townshend indulged an enduring fascination for turnips by inventing a crop-rotation scheme that included turnips in one rotation. He was, in effect, simply reinventing the old Saxon three-field crop rotation, but his system involved dividing the land into four fields. The cycle involved first planting with wheat, then with clover, followed by barley and finally, of course, turnips. Like the peas and beans of the Saxon system, clover is a legume and replenishes the soil with nitrogen. Turnips and clover are also fodder crops, so could be fed to livestock. This successful arable rotation earned Charles the unfortunate nickname of 'Turnip Townshend' – and nobody named a progressive rock band after him.

Beginning in the seventeenth century, a series of Inclosure Acts took land out of common ownership and transferred it to individual landowners. Between 1604 and 1914, over 5,000 individual Acts enclosed over 7 million acres of land and produced the outlines of the countryside familiar today. Individual ownership meant that farming was now embedded firmly within the profit-driven market economy. This encouraged the enclosure of more and more land to maximise profits, including the so-called wastes – moors, heaths and marshes that had become the last refuges for our beleaguered wildlife. Improving technology allowed even these marginal lands to be brought into productive – and profitable – cultivation.

These changes didn't pass without note. John Clare (1793–1864), a farmer, countryman and poet, was appalled by the changes wrought by enclosures. In 1821, in the poem 'The Village Minstrel', he lamented:

There once were lanes in nature's freedom dropt,
There once were paths that every valley wound –
Inclosure came and every path was stopt:
Each tyrant fix'd his signs where paths were found
To hint a trespass now who'd crossed the ground.

Clare would have been even more appalled by the changes since his death. He'd already seen the rise of powered farm machinery in the early nineteenth century, along with the arrival of railways that, following the pioneering line running between Stockton and Darlington, had sprawled across the land, allowing farm produce to speed its way to an ever-expanding urban market. Early in the twentieth century, the first pesticides became available and began the process of denuding the countryside of even the tiniest creatures.

At the same time, a process for making artificial fertilisers was invented, so we also became less and less reliant on crop rotations, which had, at least, created some diversity in the countryside. New strains of plants increased yields and, together, these factors began to create vast tracts of monoculture with little room for wildlife. In the late nineteenth century, a farmer could expect three-quarters of a ton of wheat from every acre of land. A modern farmer can grow over three tons per acre.

With a burgeoning human population, it can certainly be argued that these increases in productivity are a necessity, and Britain's farmers can be rightly proud of their achievements. But first, these advances have come at a huge environmental cost – which indicates the urgency of reassessing how agriculture works as part of a new mindset for the twenty-first century. Second, if we view this in a worldwide context, the need for a serious revision is even more stark. Globally, we already produce enough food to feed 11 billion people. In 2024, there were 8.1 billion people on the planet, of whom 825 million didn't have enough food. Over 10 million children die each year from starvation or related diseases.[*] Yet at least a fifth of all food produced globally is wasted.[†] Business as usual is not just inadvisable, it's immoral.

We've seen over the last few pages how much the landscape has changed over the millennia since the ice retreated from our shores – slowly at first, then gathering pace. And all the time, the natural world has been pushed further to the fringes. But the pace of change in the last half of the twentieth century was beyond anything that came before, beginning with the 'Dig for Victory' campaign in the Second World War. Hay meadows were replaced with intensively fertilised grass leys, resulting in the loss of 97 per cent of these flower- and insect-rich grasslands. Between 1930 and 1984, that translates to 7 million acres. Hedgerows, many planted to demarcate private fields during the enclosure period, were grubbed out. In 1946, about half a million miles of hedgerows gave the British countryside its character. By 1993, half of these had gone, and with them homes for many farmland birds and insects. Nearly half of our butterfly species breed in hedgerows.

Today, our green and pleasant land holds some of the lowest natural diversity and abundance anywhere on the planet, not an accolade to be proud of. Three-quarters of our land is farmed, and much of that is a desert as far as wildlife is concerned. Of course, we need to produce food, and I certainly

---

[*] Mark A. Maslin. *How to Save Our Planet: The Facts.* Penguin. 2021.
[†] World Food Programme, 2024. https://www.wfp.org/stories/5-facts-about-food-waste-and-hunger

don't blame modern farmers for the parlous state of British wildlife. Many of them are deeply concerned with balancing the needs of production with those of a diverse natural world. They are, though, like the rest of us, enmeshed in economic and political systems that make no sense whatsoever. Indeed, many farmers today find it increasingly difficult to make any kind of living at all. In earlier chapters, I lamented the woeful lack of joined-up thinking and the deeply flawed economic models that have made life so difficult – for people as well as wildlife. A similarly short-sighted approach has also shaped the farming landscapes of Britain. However, we'll have more to say on this when we reach the river's source – where, from the top of Cross Fell, we have an overview of how all the threads of natural, social, political and economic history along the river can be woven together and how this perspective might guide us in the future. First, though, we have to complete our journey. It's time to head upstream once more.

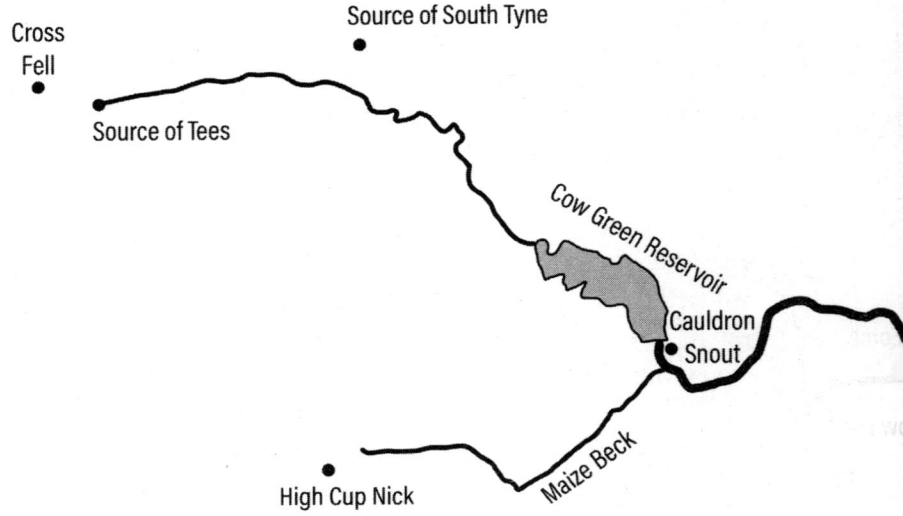

Cross Fell

Source of South Tyne

Source of Tees

Cow Green Reservoir

Cauldron Snout

High Cup Nick

Maize Beck

River Lu

River Balde

1 km

PART III

Upper Reaches

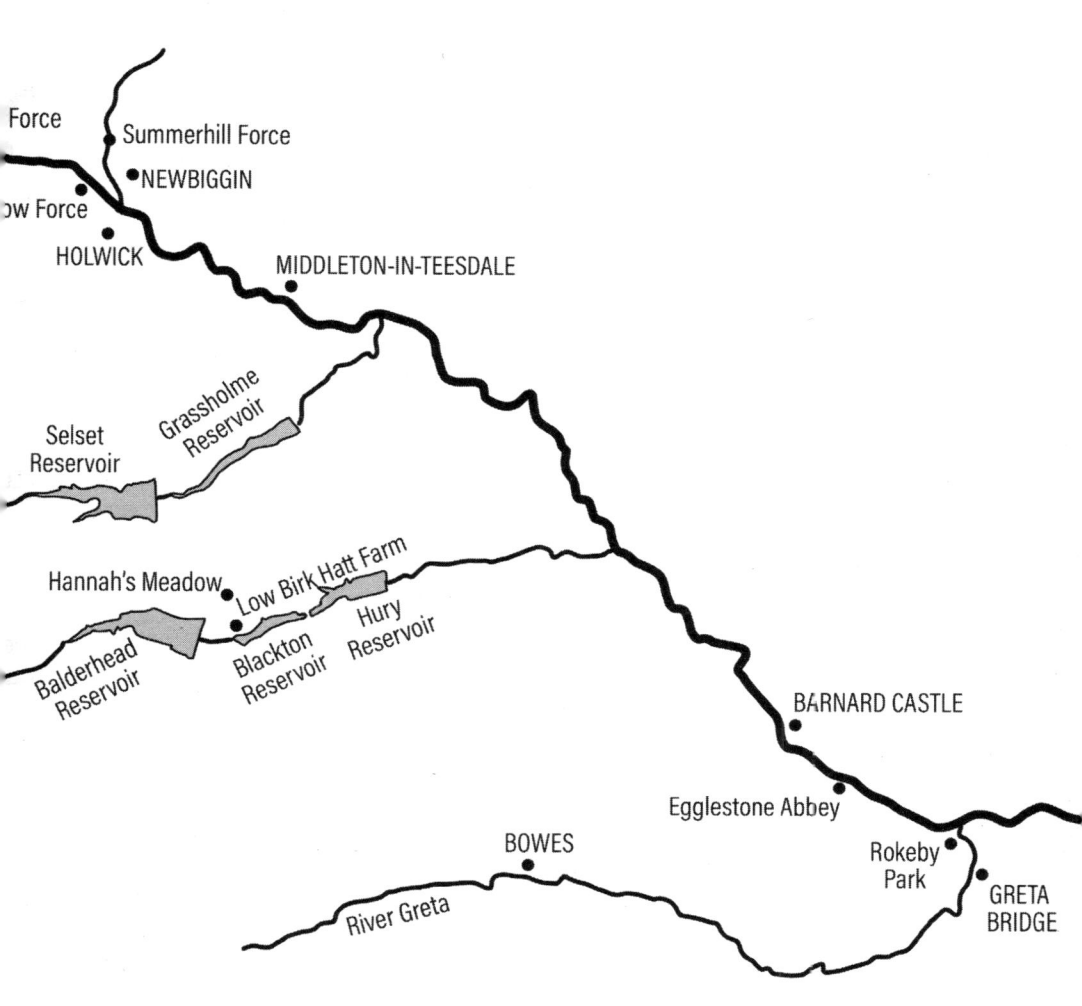

Force
Summerhill Force
●NEWBIGGIN
ɔw Force
HOLWICK
MIDDLETON-IN-TEESDALE
Grassholme Reservoir
Selset Reservoir
Hannah's Meadow
Low Birk Hatt Farm
Hury Reservoir
Balderhead Reservoir
Blackton Reservoir
BARNARD CASTLE
Egglestone Abbey
Rokeby Park
GRETA BRIDGE
BOWES
River Greta

8

Gateway to the High Country

*Rokeby to Middleton-in-Teesdale – Dippers,
Dickens and Dalesfolk*

> It's my favourite place, here… I stand here and watch
> the seasons come and go. At night, the moonlight plays
> on Hunder Beck… and the waters sing a song to me.
> If I have nothing in my pocket, I will always have this.

DESCRIPTION OF LOW BIRK HATT FARM, HANNAH HAUXWELL
(and Barry Cockcroft), *Seasons of My Life: Tales from
a Solitary Daleswoman*, 1989

> A nobler want of man is served by nature,
> namely, the love of Beauty.

Ralph Waldo Emerson,
Nature, 1836

In the 1720s, as part of a tour of Britain, the novelist and journalist Daniel Defoe visited the mountains of northern England, only to denounce them as 'the wildest, most barren and frightful of any that I have passed over'. Worse still, they had 'no lead mines and veins of rich ore, no Coal Pits... of no use or advantage to Man or Beast'.

A century later, as the environmental and social damage caused by lead mines and coal pits – by industrial capitalism in all its forms – was becoming clear, attitudes to wild places such as the upper Tees had changed. In the middle of the nineteenth century, the artist Alfred William Hunt, recently returned from the continent, said of the River Greta, 'all Germany and Switzerland put together would not be worth a mile or two of Greta from point of view of delightfulness'. High praise indeed, if probably a little exaggerated. Then again, maybe not. I am walking this stretch of the river on a cold but bright January day. Sheets of ice have grown over the water in exotic and flamboyant swirls, creating abstract works of art stretching between the boulders at the river's edge. And as I clamber up through steep woodlands on the bank of the Greta by the Meeting of the Waters, I come across great clumps of snowdrops, glowing in winter sunshine and set against the sparkling river below. Perhaps Alfred William Hunt had it right after all.

From the Meeting of the Waters and on through Barnard Castle, we enter a world very different from the gentle agricultural land of the Middle Tees – that of the dales – Deepdale, Baldersdale and Lunedale – the major side-valleys of the upper Tees, carved by the tributaries Deepdale Beck and the Rivers Lune and Balder, which run from the high moors to the south of the Tees. Hike far enough along these dales and you'll find yourself in some of the most remote areas in England. In winter, they can be bleak – and dangerous. Even local people, well versed in the dark moods of the dales and high moors at this time of year, have died in arctic whiteouts. Yet in summer, the moors and meadows resound to the calls of curlews, golden plovers and lapwings, the music of the high country. In spring, meadows in the valleys are painted gold with buttercups and, later in summer, spangled with the magenta disks of wood cranesbill and the frothy-white flowers of meadowsweet.

Nature has been in retreat across most of our rural land-scapes over many centuries, although more dramatically in the last few decades. In Teesdale, though, there are still places where we can step back in time, far enough to appreciate at least a little of what has been lost. Later, we'll explore such a place in Baldersdale, once the home of the remarkable woman whose words open this chapter, but first we have to get there. Baldersdale lies beyond Barnard Castle and between where I stand now, at the confluence of the Tees and the Greta, and the confluence of the Tees and the Balder, there's a lot of natural and human history to explore.

Heading upstream from the Meeting of the Waters, the path follows the southern bank of the Tees, which here is flanked by steep wooded banks. It's still early in the year and oak and ash have yet to leaf out, so I have a clear view of the rushing, shallow river as it bubbles and cascades over riffles, created where its stony bed rises close to the surface. A short, raspy *zheet zheet*, barely audible over the babble of water, warns me that I've entered dipper territory. If any bird is a symbol of the wilder upland stretches of a river, it's this little ball of feathers. That's a flight call, and it draws my attention to a bird flying fast and low on short, whirring wings along the river. But it'll be a lot harder to spot when it lands on a stone in mid-river. With a dark-brown back and belly and a white throat, and its characteristic habit of bobbing up and down, it all but disappears among the tumbling, foaming water.

Before I spot it, I hear it again. This time it's singing – a tune that babbles and cascades like the river that it calls home. The songs of dippers are often hard to hear above the noise of a river – how on Earth they ever hear each other, I'll never know – but fortunately I've reached a smoother, quieter stretch of water and the song allows me to pinpoint the singer. He is perched on a small boulder in the centre of the river, about sixty feet upstream, and he's so focused on his performance that he doesn't seem to be paying me much attention. So, I nestle down next to a large boulder on the riverbank to watch.

He sings on and off for about ten minutes, and then hunger gets the better of him. He starts by plunging his head into the water around the boulder that served as his stage. When he stands up, I can see some kind of insect nymph dangling briefly from his bill before he swallows it. It's quite a hearty meal, so it may be a stonefly nymph, some of which reach an inch or more in length, but he's too far away for me to be certain. Dippers frequently submerge themselves completely when searching for food and often have a favourite stone from which to launch their underwater forays. This one flies upstream, even closer to my half-concealed position, and lands on another small boulder. Facing upstream, he makes a little arching jump and plunges below the surface. About ten seconds later, he bobs up again and scrambles back onto his boulder, a large prize clamped in his beak. Through my binoculars, it looks as though this time he's got a caddis fly larva, sheathed in a stony case. A deft flick of his head and he flips the case off. In almost the same move, the exposed juicy larva is swallowed. He's obviously done this before. This seems to be one of his favourite hunting spots since, with almost no pause, he disappears again beneath the roiling water.

Dippers have been one of my favourite birds ever since I was a teenager, when Ken Smith showed me my first one, on a small stream near Kildale at the edge of the North York Moors.

These dumpy little birds always struck me as ill-adapted to the pursuit of prey underwater in fast-flowing rivers so, even after my very first encounter, I was curious about how they hunted so effectively beneath the surface. Many years later, I got the chance to find out.

In the 1980s, for a series of films on British wildlife, we sank a remotely operated camera in an underwater housing into a river frequented by dippers. After watching one bird for long enough to work out where its favourite spot was, we bedded the camera into the gravel riverbed close by. Cables ran back to the shore, one to a monitor which allowed us to see what the camera was seeing, and another which could run the camera should the dipper happen to feed in just the right spot. It took a few days, but we eventually recorded several shots of a dipper feeding underwater – a privileged glimpse into a part of the life of a dipper not usually visible to birdwatchers.

To reach the riverbed, it used its wings to 'fly' down through the turbulent water. Once there, it used its powerful feet to grip stones as it walked upstream, often with wings half-spread – perhaps using them as a submarine uses angled hydroplanes to keep it submerged. It rooted around in the crevices between smaller stones and in just a few seconds it had an insect larva clamped in its beak. All it then had to do was to let go, and it bobbed back to the surface like a cork.

Dippers do have their limits, though. As we've seen, the Tees can change from a tranquil river to a monster in full spate without warning, especially when heavy rain or snowmelt in the high fells creates a Tees roll. Sensibly, dippers avoid feeding underwater when rivers are in flood. One study found that under such conditions they find safe perches, often under bridges, and wait until the river falls to more manageable levels, although, if they get peckish, they can still catch insect prey by scrabbling around stones at the river's edge, and they can even catch insects on the wing.

Another study suggests that dippers might have to hone their aerial skills even more as we change their world. In south-western Ireland, detailed measurements of dippers have shown that their legs are getting shorter, while their wings and tails are getting longer. Natural selection is shaping the birds to cope with lower volumes of water in the rivers as global warming reduces rainfall. Instead, longer wings and tails give them more aerial agility as they come to depend more and more on flying insects.

Evolution is usually seen as a process that unfolds imperceptibly over long spans of time, yet natural selection is continually tinkering with both bodies and behaviour, generation by generation, and when the selective pressure is strong, those changes may happen quickly. We've altered the planet so much, and in the last century so quickly, that we've created strong selective pressure on all the organisms that share our lives. Think of antibiotic-resistant bacteria or pesticide-resistant pests. The same forces are shaping all kinds of creatures in surprising ways. On the winter marshes of Saltholme, we've already explored the behavioural and ecological changes that have been caused by the widespread provision of winter food for birds, but our impacts on nature are so pervasive that we're also causing fundamental changes to the DNA of many creatures in the same way that natural selection has done over the last 3.5 billion years.

For my 2013 documentary *Unnatural Selection*, which I introduced while we were watching flocks of birds on feeders on the North Tees Marshes, I wanted to highlight this almost unknown problem for conservationists and environmentalists. We've already met the finches on the Galapagos Islands, whose evolution has been thrown into reverse by easy access to discarded human food, but the problem is a global one. For this reason, I travelled across Europe, North America and Australia, all places where these unnatural pressures are often greatest, and found some very unexpected stories. In Nebraska, I met with scientists who'd been studying the local cliff swallows for thirty years, particularly those that nest under bridges along freeways. They'd discovered that the wings of these swallows, unlike those of the dippers, had grown shorter. Over the same three decades, deaths from collisions with trucks had decreased, leading the scientists to conclude that shorter wings made the swallows more manoeuvrable and better able to avoid onrushing sixteen-wheelers. Of course, there may also have been changes in behaviour – freeway swallows might have become more traffic-savvy – but in either case we've affected the course of their evolution.

I found similar stories repeated across the world. In Australia, in areas infested with introduced and very toxic cane toads, snakes are evolving smaller mouths. Those with smaller mouths are unable to eat cane toads and so survive better than those that are able to feast on this abundant but deadly prey, thus spreading the genes for smaller mouth sizes through the whole population. In Chesapeake Bay in Maryland, local diamondback turtles are evolving larger adult sizes. Smaller adults crawl into lobster pots, from which they can't then escape. Since they must breathe air, these elegantly spotted turtles soon drown, eliminating genes for smaller adult sizes from the population.

In my walk along the Tees from Greta Bridge to Barnard Castle, I've seen plenty of much more obvious examples of such unnatural selection – in the varied shapes and sizes of the dogs I've met along the path. Every one of them, from yapping little Yorkshire terriers to golden retrievers, plunging into the freezing river with careless abandon, is descended from the wolf in the distant past but, over the millennia, we've selected for different traits in different breeds. The sheep and cows in the fields beside the river are other familiar examples. Such unnatural selection is the way we've domesticated a large variety of plants and animals. The point is that this is deliberate and planned – but the scale of all the unnoticed, accidental effects is truly astounding. Our impact on the planet is so ubiquitous, it's even redirecting evolution itself.

For the moment, the waters of the Tees still flow abundantly and the dippers I'm watching are catching their fill beneath the surface. Still crouched beside the riverside boulder, I'm roused from my contemplations when a much larger bird hurtles by, heading upriver before performing a handbrake turn into the trees on the steep bank opposite. A goosander. It's a duck, though not the kind you would ever see paddling around a park lake. Goosanders are birds of wild, swiftly flowing rivers. Not that long ago, I would've had to make a trip much further north in Europe to see these intriguing birds nesting, but in the second half of the nineteenth century, they began to extend their breeding range south.

They first nested in Scotland in 1871, although there's some suggestion that they may have nested earlier than this on South Uist in the Outer Hebrides. By 1941, they'd begun nesting in the North-East of England, and by the 1980s, they'd reached the wild rivers of Devon, flowing off Dartmoor. This dramatic expansion was mirrored by a close relative, the red-breasted merganser. It's quite hard to tell the russet-headed females of these two species apart without a clear view, but the males couldn't be more different. Male mergansers sport a spiky punk hairdo that sprouts from their iridescent-green heads. Goosander males have similar dark-green heads but are more clean-cut, with a bouffant style. In addition, the flanks of male mergansers are patterned with a fine filigree of black and white, while those of goosanders are pure white.

Both are sawbills, as are the smews that we met on Teesmouth, and all three ducks feed on small fish caught with their serrated bills after fast underwater chases. They nest in holes in trees bordering their riverine fishing grounds, although goosanders favour wider, slower sections of rivers than mergansers. As breeding birds, mergansers are also less widely distributed. They're more abundant along the west coast of Scotland and don't, as yet, nest along the Tees. So, I'm pretty sure the bird I saw – a female – was a goosander, even though it was there and gone in a flash of whirring wings.

Presumably, somewhere along the river is her mate, but not for much longer. Like many ducks, goosanders undertake a 'moult migration' – they head to places where they can find both safety and ample food during the time-consuming and energy-sapping process of replacing their feathers. It's especially important for ducks since they replace all their flight feathers at the same time, which leaves them flightless for several critical weeks. We've already seen that many of the shelducks from the Tees Estuary, along with those from most of the rest of Europe, travel to the German part of the Wadden Sea to moult. Goosanders from Britain make an even more impressive journey.

The males fly all the way to North Cape at the very top end of Norway. It's a journey of over a thousand miles, and to arrive in the far north in time to moult, they must leave the rivers – and their mates – during May. But before they set off on their epic trek, they first congregate at gathering sites. One such place is Hoselaw Loch in Scotland (although it's only just in Scotland; it's a small lake just a couple of miles over the border from Northumberland). Goosanders ringed on the Tyne turn up here, where presumably they're joined by those from the Tees, including the mate of the one I've just seen heading off into the woods.

The females are left behind with all the family duties. When the males leave, the females are already sitting on perhaps a dozen or so eggs, which by summer will have hatched into a dozen hungry ducklings. By the time the ducklings can fly and look after themselves, it's far too late in the year for the females to make the long journey north. They'll have to moult much closer to home.

Meanwhile, on the estuary of Norway's Tana River, where it empties into the Barents Sea at North Cape, around 35,000 male goosanders from Britain and Western Europe are loafing on the water. There are isolated sandbanks in the estuary where, like the moulting shelducks on the Wadden Sea, the goosanders are safe from predators. The waters around North Cape also abound in sand eels, which provide plentiful food for the ducks as they replace their old feathers. Even so, it seems like an inordinate effort to reach this isolated spot at the very top end of Europe.

On the Tees, the female goosander is in for a busy summer. Somewhere in the woods beyond where I'm sitting is her nest, a hole high in a large tree. Back on the summer marshes of Teesmouth we met burrow-nesting shelducks, but in Britain several other species nest in holes, especially in tall trees. Apart from goosanders and red-breasted mergansers, these include native (though rare) goldeneyes and introduced mandarin ducks.

I've filmed mandarin ducks in the New Forest in Hampshire as well as the related Carolina wood duck in North American forests, and managed to capture the moment that the eggs hatch and the ducklings are forced to leave their high-rise residence. Their mother flies to the ground just below the cavity and calls to her new family. The chicks all reply and, with much cheeping, clamber up to the edge of the hole to peer out, seemingly incredulous at finding themselves perhaps thirty feet above the ground. It's hard not to imagine them thinking, 'Why have you done this to us, Mum?' They won't be able to fly for many weeks, so their only option is to jump – if they can pluck up the courage. They balance on the rim of their hole full of bold intentions, only to jump back into the safety of the nest. Eventually, the bravest duckling takes a leap into the unknown. With stubby little wings and webbed feet spread wide to act as a parachute, the chick plummets to the ground. The little ball of fluff is so light and so well cushioned by a thick layer of down that it bounces when it hits the forest floor and tumbles head over feet a few times before righting itself. Cheeping loudly in exultation, it runs full-pelt towards its mother.

Not all the chicks are as brave as the first and it takes time for the whole family to assemble on the forest floor. At long last, their mother leads a long line of noisy chicks through the forest to water, which, for the wood ducks I filmed, was several hundred yards away. I've never found a goosander's nest, but I imagine a similar scene will play out in a few weeks' time, with the chicks bouncing and rolling down the steep slopes to the Tees. These moments are among the most charming of any in the natural world, but the arrival of a new brood of goosanders on the Tees doesn't please everyone.

Goosanders are fish-eaters and young salmon are an ideal size for these birds. There are frequent calls to cull goosanders (and cormorants), especially on salmon rivers, although the evidence against them is, to say the least, equivocal. Detailed studies in British Columbia in western Canada, where goosanders are known as common mergansers, showed that culling had no effect on Pacific salmon numbers, since here the goosanders fed on other kinds of fish, including several predatory kinds. On the eastern coast of Canada, goosanders do eat the smolts of Atlantic salmon, but reducing the goosander population to as few as three birds per fifteen miles of river had no effect on salmon numbers. A significant increase in salmon numbers was only observed at even lower goosander densities than this – in other words, by all but wiping out these ducks completely.

A goosander's diet varies widely in both time and space, so it's hard to apply findings in any one place to somewhere else. But, in my mind, even this is not the point. It's another example of the problem of managing nature purely for human benefit. And even if we do accept this premise, why is my right to enjoy goosanders outweighed by another person's right to enjoy hauling salmon out of the same rivers? Well, one answer of course is money. People pay lots of it for the privilege of catching salmon, whereas I pay nothing to sit quietly by the river and watch its birds. But is this the most sensible criterion on which to base our national and local conservation strategies? As we've seen many times in our journey, a pure market economy hasn't done a great job of protecting our biodiversity. Instead, it's been a major factor in reducing Britain to its current nature-barren status. Controversy over goosanders on salmon rivers rarely makes the headlines, in comparison, for example, to the parallel story of the illegal killing of hen harriers on grouse-shooting moors. But both stories raise fundamental issues about who the management of our countryside is for. In any case, recent figures suggest that while there are about 3 million anglers in Britain, there are as many as 6 million birdwatchers.

In addition, it's not goosanders – or mergansers or cormorants – that have reduced Atlantic salmon to such low numbers that many scientists predict their imminent extinction in English rivers in perhaps a matter of decades. Their precarious state is down to climate change and pollution in both rivers and the oceans, along with pests and diseases from salmon farms. River pollution comes from the runoff from agricultural land soaked in fertilisers, herbicides and pesticides, and the appalling amount of raw sewage that the private water companies are still dumping into our rivers. Salmon in the Tees have also had to contend with the ill-considered construction of barriers to migration, such as the Tees Barrage. It's the same complete lack of an integrated approach that we've encountered many times in our journey along the Tees.

As if to make the point, the female goosander reappears and is gone in an instant. I wish her good fortune with her family in the summer ahead and rise, somewhat stiffly, from my crouched position beside the mossy boulder. Time to stretch my legs and head into Barnard Castle to find a much-needed coffee. In a short time, I see an old, crenellated bridge spanning the rocky banks of the river and the path soon ascends a short, steep slope to emerge from the woods and onto the bridge. Standing in the middle, I gain a lovely view of the Tees, at this point flowing through a low, rocky gorge topped by a tangle of trees. This is Egglestone Abbey Bridge, and a short walk along a surprisingly busy minor road brings me to the ruins of Egglestone Abbey itself.

I can both see and hear it as I walk up the short path to the ruins. It resounds with the *kyaa-kyaa* of about a dozen jackdaws, deep in conversation among themselves. They've probably found some secure holes in the remaining stonework to build their straggly nests. Like rooks, they nest communally, although jackdaw colonies are often just small villages compared to the busy metropolis of a large rookery.

Apart from the jackdaws, the place is deserted. It was only ever a small abbey and there are just a few walls left standing. Some of the stonework was even torn down to pave the stable yard at Rokeby Hall. So, while it doesn't compare to the spectacular great abbeys of North Yorkshire – Ripon, Rievaulx, Fountains, Whitby and others – it's still a fascinating place. Founded in the late twelfth century, it was home to the White Canons, monks of the Premonstratensian order, originally founded at Prémontré in northern France. They gained their name from the white habits they wore and, although it was a strict order, the White Canons were frequently out and about in the local communities, distributing meat and drink to the poor.

The abbey itself was also poor, often barely able to support itself, a situation not helped by its location. The site was originally chosen because it had good access to the Tees and was surrounded by rich farmland. But it was also in a disputed border zone and was ransacked in 1315 by a Scottish army. A few decades later, it suffered a similar fate, this time at the hands of an English army marching north to meet the Scots during the Second War of Scottish Independence.

It was probably of little consolation to the White Canons that in 1346 an English army, under Lord Neville, won a decisive victory at the Battle of Neville's Cross, just outside Durham, capturing King David II of Scotland and routing an army that was double the size of the English force. Despite such setbacks, Egglestone Abbey struggled on until King Henry VIII dealt a final killer blow to all abbeys, large or small. After the Reformation and Dissolution of the Monasteries, between 1536 and 1541, Egglestone Abbey passed between several owners until, in 1770, it was bought by John Morritt of Rokeby Hall, who then helped himself to some of the stones for construction at the hall just downriver. His descendants handed it over to the nation in 1925. Today, it's under the watchful eyes of English Heritage and about a dozen jackdaws.

It's not far from Egglestone Abbey to Startforth on the southern bank of the river. Here, a road bridge crosses the Tees, overlooked by a steep cliff topped by the ramparts of Bernard's castle, which gives the town of Barnard Castle its name. Although the castle now lies in ruins, it still occupies an imposing position and must have been a formidable sight in former days. Walter Scott's epic poem *Rokeby* opens on a blustery night, probably not far from where I'm now standing:

> The Moon is in her summer glow,
> But hoarse and high the breezes blow,
> And, racking o'er her face, the cloud
> Varies the tincture of her shroud;
> On Barnard's towers, and Tees's stream,
> She changes as a guilty dream,
> When Conscience, with remorse and fear,
> Goads sleeping Fancy's wild career.

Before 1066, Barnard Castle, known affectionately as 'Barney' by the locals, was included in an Anglo-Norse estate centred on the village of Gainford. It was under the control of the earl of Northumbria and a powerful bishop of Durham, the latter office descended from the highly respected bishops of Lindisfarne, whose numbers included Saint Cuthbert. After the Norman Conquest, King William I tried to gain some measure of control over this distant part of his new kingdom, although at first without much success. As we saw in Chapter 5, his attempt to install a Norman earl in Northumbria in 1080 met with disaster when the nobleman and his retinue of several hundred were all slaughtered at Durham, one of several acts of Northern independent-mindedness that would lead to the savage retaliation known today as the 'Harrying of the North'.

He obviously didn't harry hard enough. There were more rebellions in 1095, which led to further attempts, this time by William's successor, his son William Rufus (King William II), to break up the old kingdom. As part of his scheming, he gave the lordship of Gainford to Guy de Balliol. A castle of sorts was constructed upstream from Gainford, at a more easily defendable site on the natural ramparts of the cliffs that overlook the Tees at what later would become Barnard Castle. In the second half of the twelfth century, Bernard de Balliol built the more substantial stone structure that gave the town that sprang up around the castle its name.

Not long after, in 1263, one of Bernard's descendants, John Balliol, founded Balliol College in Oxford, and his son, another John, went on to become king of Scotland. Definitely a family of high achievers – not that any family connections to old Barney made the Scots think twice about raiding the Tees Valley. John Balliol's reign as king of Scotland was brief (1292–96), largely because he was nothing more than a vassal of King Edward I, the Hammer of the Scots; it wasn't an association to inspire much confidence in the cause of Scottish independence. After Balliol was deposed, Scotland was without a monarch until 1306, when Robert the Bruce came to the throne – a descendant of yet another family originally given estates along the Tees.

Barnard Castle was later inherited by King Richard III and after his death it passed into the ownership of several different families, until it was finally ruined during the English Civil War (1642–51), at the time that Scott's *Rokeby* was set. But the town of Barnard Castle continued to play a role in British literary history.

In a cold and snowy February in 1838, two young men alighted from the mail coach in Greta Bridge, a small village straddling the River Greta a little upstream from the Meeting of the Waters. They'd begun their journey two days earlier in London, transferring to the Glasgow mail coach in Grantham. On its way to Scotland, this coach crossed the Pennines through the Stainmore Pass, one of the few routes through these northern uplands, though even this is not always passable in winter. Today, it's the route followed by the A66, which is still one of the first roads to be closed when snow hits the North Country. The route at times runs close to the Tees and passes just to the south of Rokeby Park, crossing the River Greta at Greta Bridge. In February 1838, the journey was cold and arduous, but the Glasgow mail made it to Greta Bridge without incident.

The two young men on board were travelling under aliases, although the fame of one of these men was spreading so rapidly that it's unlikely that he remained incognito for long. His name was Charles Dickens and, at twenty-five, he had just had considerable success with his first novel, *The Pickwick Papers*. His travelling companion was his illustrator, Hablot Knight Browne. After an uncomfortable journey, Dickens wasn't at first enamoured of the prospect that greeted them as they alighted in Greta Bridge to find their lodgings.

> We reached a bare place with a house standing alone in the midst of a dreary moor which the guard informed us was Greta Bridge. It was fearfully cold and there were no outward signs of anyone being up at the house.

He soon changed his tune, though, when he was met with some good old Northern hospitality at the New Inn.[*]

> But to our great joy we discovered a comfortable room with drawn curtains and a most blazing fire. In half an hour, they gave us a smoking supper and a bottle of mulled port.

Dickens had begun his literary career as a political journalist, and he never quite lost his crusading stance as he used his popular novels to highlight the inequalities and injustices of the Victorian era. He had travelled to this remote corner of England to investigate reports he'd heard in London concerning some

---

[*] The building that they stayed in is still standing, although it is now part of Thorpe Farm holiday camp.

particularly brutal Yorkshire schools. A dozen such boarding schools clustered around the town of Bowes, further up the River Greta, about six miles to the west of Greta Bridge. Many of these schools advertised in London, proclaiming their policies of no vacations. This, and their remote location in the northern Pennines, made them the perfect places to dump unwanted children.

Dickens had armed himself with a cover story – he claimed to be seeking an appropriate school for the son of a friend – and intended to use this ruse to gain access to the schools and to see for himself what was going on. Before setting out, however, the two men enjoyed what can only be described as a hearty Yorkshire breakfast.

> We have had for our breakfast, toast, cakes, a Yorkshire pie, a piece of beef about the size and much the shape of my portmanteau, tea, coffee, ham and eggs... we start in a postchaise for Barnard Castle, which is only four miles off.

In Barnard Castle, they stayed in the King's Head, which stood until recently just above the Buttermarket in the town's centre. Here, they could make inquiries as to those schools with the worst reputation. Not far from their inn, Dickens discovered an intriguing clock shop owned by Thomas Humphrey, which included among its inventory an impressive timepiece, described as a 'centre-second pendulum clock, dead-beat movement, possessing a compensation pendulum for change of temperature'. The clock, Thomas explained, had been made by his son, William. Dickens was so taken with its craftmanship that the clock became the inspiration for his weekly periodical, *Master Humphrey's Clock*, which published many short stories as well as two of his novels, *The Old Curiosity Shop* and *Barnaby Rudge*. Thomas Humphrey also drew Dickens's attention to the Bowes Academy, run by one William Shaw, who was said to mete out particularly harsh treatment to the boys.

Dickens heard similar reports from others in Barnard Castle, so he and Browne set off to Bowes to pay headmaster Shaw a visit.

Although Dickens and Browne are known to have met Shaw, it's not clear how long they spent with him or how much of the school they saw for themselves. However, the encounter was enough to inspire Dickens to write the novel *Nicholas Nickleby*, published in 1839. The school at Bowes became Dotheboys Hall and William Shaw was caricatured as the headmaster Wackford Squeers, although, perhaps to avoid litigation, Dickens always maintained that the character of Squeers was an amalgam of several different people.

Nicholas Nickleby dramatically evokes the horrendous lives of the boys at Dotheboys Hall, but almost as soon as the book was published, it engendered lively debate as to its accuracy. The Yorkshire schools were certainly tough places. The parish register at Bowes records a long list of deaths of boys from all the schools in the area, but it doesn't suggest that the boys at Mr Shaw's school fared any worse than those at any of the other establishments.

However, while some former pupils interviewed later in their lives remembered harsh and cruel treatment, others recalled their time at the Yorkshire schools if not with fondness, then with far less horror. More recently, William Shaw's great-great-grandson, Ted Shaw, has tried to paint a more balanced picture of his ancestor. In 1988, to celebrate the 150th anniversary of Dickens's visit to Teesdale, Ted Shaw met with Christopher Dickens, Charles's great-great-grandson, an encounter that proved more congenial than the master novelist's original meeting with the headmaster in 1838, since both men were united in their praise of Dickens as an author who could elicit a deep sympathy with those unfortunates living in the dark underbelly of Victorian society. Whatever the truth of the situation in and around Bowes, the power of the pen, especially one wielded by Charles Dickens, proved overwhelming. The outcry provoked on the publication of *Nicholas Nickleby* resulted in the closure of many of the Yorkshire schools. However, the Bowes Academy building still stands on The Street in Bowes, just a few yards from the busy A66. It's a lot less remote now than it was in Dickens's day.

Not far from Bowes Academy, hidden down a small side-street, stand the remnants of the medieval Bowes Castle. It was built on the site of a much earlier Roman fort called Lavatrae, which defended a vital Roman road through the Stainmore Pass across the Pennines. This route has been a crucial link between east and west and between England and Scotland for at least 2,000 years. Looking west from Bowes Castle today, the A66 road snakes off into the distance, flanked by imposing high ground to the north and south. From the grounds of Bowes Castle, it's easy to grasp the geographical, historical and strategic importance of the Stainmore Pass. An earlier timber castle stood on this site, but a more secure stone castle was built beginning in 1170, on the orders of King Henry II, to defend against the ever-present threat from the Scots allied with the rebellious Northerners.

The Bowes family, from whom the late Queen Mother, was descended,* may have originally taken their name from the castle and the town that grew up beside it. One suggestion is that 'Bowes' may have signified the head of a regiment of Norman archers garrisoned here to guard the Stainmore Pass, although the original name of the place – Boga – from which the modern name derives was just as likely to refer to a bow or bend in the river as to a weapon.

In 1310, Sir Adam Bowes married Alice Trayne, heiress of Streatlam Castle in County Durham, and this remained the family seat until the nineteenth century, ending any association with the village of Bowes. The family did, however, retain strong connections with Barnard Castle. For generations, as stewards of the castle, they defended the ever-changing border country between England and Scotland, although things got a

---

* In more recent times, the family name changed from Bowes to Bowes-Lyon.

little out of hand in 1569. The steward of Barnard Castle at the time was George Bowes, who found himself embroiled in the Rising of the North, when, yet again, the Northern earls were in rebellious mood.

This time, they were backing attempts to replace Protestant Queen Elizabeth I with Catholic Mary, Queen of Scots, a plot that was hatched at Raby Castle further down the Tees. George Bowes remained loyal to Elizabeth and the Protestant cause but, at Streatlam, he found himself surrounded by a lot of people who didn't share his views. Sensibly, he took refuge behind the stout stone walls of Barnard Castle. However, the rebels besieged the castle and forced George Bowes to surrender after eleven harrowing days. Ultimately, the plot failed, and George Bowes's loyalty was rewarded with the lease of lead mines, which were becoming an increasingly important part of the economy of the northern Pennines.

With interests in coal mining as well, the Bowes family grew wealthy. In less uncertain times, a descendent of George, John Bowes (1811–85), spent huge sums on assembling an extravagant art collection. He also ran a successful stud at Streatlam, where he bred four Derby winners in twenty years, while one of the horses, West Australian, was the first to win the Triple Crown.* By the middle of the twentieth century, though, Streatlam Castle was proving too expensive to maintain so, in 1959, it was blown up with dynamite in a training exercise by the Territorial Army. That's one way of limiting liabilities. The extensive art collection can, however, still be seen at Barnard Castle. Although many pieces made their way to the Metropolitan Museum in New York, much of the collection is on display in a purpose-built museum on the eastern edge of town.

In 1852, John Bowes married the French actress and fellow art collector Joséphine Benoîte Coffin-Chevallier and together

---

* The Triple Crown in horse racing denotes victory in the 2000 Guineas (at Newmarket), the Epsom Derby and the St Leger (at Doncaster).

they commissioned a massive building, built in the style of a grand French palace, to house their collection. The Bowes Museum, grandiose to the point of being a monstrosity, opened in 1892, after both John Bowes and his wife had died. Although it houses some important artworks, it exudes such opulence and decadence that I find the place both intriguing and uncomfortable in equal measure. Perhaps this is because, according to an article from the *Metropolitan Museum Journal*, the objects were acquired 'as part of a pattern of luxury spending and social display by a family determined to make its mark in Georgian society'.* It's a stark example of the conspicuous consumption which is continuing to drive the unsustainable use of many raw materials today.

The centrepiece of the Bowes collection is a silver swan, first recorded in a London museum in 1744, before eventually being acquired by the Bowes family in the nineteenth century. Every now and then, a clockwork mechanism animates the swan in front of an appreciative crowd, an early example of animatronics that once amused emperors and kings. Today, even in an age of photorealistic computer graphics, it still captivates crowds, as it also intrigued the usually cynical Mark Twain, who saw the swan in action at the Paris Exhibition of 1867.

I watched the Silver Swan, which had a living grace about his movement and a living intelligence in his eyes – watched him swimming about as comfortably and unconcernedly as if he had been born in a morass instead of a jeweller's shop – watched him seize a silver fish from under the water and hold up his head and go through the customary and elaborate motions of swallowing it.

Twain's description of the silver swan is taken from his book

---

* Margaret Wills and Howard Coutts. 'The Bowes Family of Streatlam Castle and Gibside and Its Collections'. *Metropolitan Museum Journal* 33, 231–43. 1998.

The Innocents Abroad (1869), which, masquerading as an innocent travel book, documents his insightful impressions of Europe and the Holy Land during his extensive travels through the area. Writing with his characteristic wit and stinging satire, Twain is frequently bemused and occasionally outraged by the adherence to the old ways and old ideas by Old World societies locked firmly in the past. Twain never visited the Tees Valley (which I'm sure he must have regretted), but another equally influential American writer, who made similar critical observations on English society, did come here – and even gave a lecture at Barnard Castle.

Ralph Waldo Emerson (1803–82) was one of the first modern environmentalists. His work has inspired many later environmental and nature writers, and, like Twain, he travelled extensively in Europe, including England. On 8 February 1848, he spoke at the Mechanics Institute in Barnard Castle, after which he took the opportunity to explore the Tees and the surrounding countryside, following roughly the same path that we've been treading along the river. In 1856, he wrote *English Traits*, based on his travels around the country, in which he comments on the gaping chasm between the rich elite, their inherited wealth, grown even fatter during the Industrial Revolution, and the poor masses on whose labour that wealth was created.

> There is no country in which so absolute a homage is paid to wealth. In America, there is a touch of shame when a man exhibits the evidences of large property, as if, after all, it needed apology. But the Englishman has pure pride in his wealth, and esteems it a final certificate ... The inequality of power and property shocks republican nerves.*

> In the country, the size of private estates is more impressive. From Barnard Castle I rode on the highway twenty-three miles

---

* Ralph Waldo Emerson. *English Traits*. Phillips, Sampson, and Company. 1858.

from High Force, a fall of the Tees, towards Darlington, past Raby Castle, through the estate of the Duke of Cleveland ... These large domains are growing larger. The great estates are absorbing the small freeholds. In 1786 the soil of England was owned by 250,000 corporations and proprietors; and in 1822, by 32,000.

Emerson's observations were an early recognition of social injustice that, as we've seen in several places along our journey, is still growing, driven by obsolete though still widely practised and destructive economic and political models. Today, land ownership in Britain remains extraordinarily unequal. About half of the country is owned by less than 1 per cent of its population. Of course, as the nineteenth century progressed, the America that Emerson admired would soon adopt those same models and acquire its own elite of the super-wealthy at the head of vast monopolies, which enabled them to exercise control over both national politics and the working population. In terms of the unequal distribution of wealth, the United States has now far surpassed the examples that Emerson saw in his travels around England.

Emerson's first major work, and the one for which he is most often remembered, was an extended essay called *Nature* (1836), in which he explored another theme that we've already touched on – and will do so again before we reach our destination. He describes our relationships with the natural world, in both practical and spiritual terms, and eloquently discusses the raw power of nature to heal the soul, while recognising that so few of us take advantage of this wellspring of wellbeing.

To speak truly, few adult persons can see nature. Most persons do not see the sun. At least they have a very superficial seeing. The sun illuminates only the eye of the man, but shines into the eye and the heart of the child. The lover of nature is he whose inward and outward senses are still truly adjusted to each other;

who has retained the spirit of infancy even into the era of manhood. His intercourse with heaven and earth, becomes part of his daily food. In the presence of nature, a wild delight runs through the man, in spite of real sorrows ... A nobler want of man is served by nature, namely, the love of Beauty.

Many naturalists and scientists that I know, even some of the most illustrious, have retained this childlike and innocent sense of wonder in nature. E. O. Wilson of Harvard University, an eminent entomologist and founding father of sociobiology and island biogeography, once admitted in a radio interview that 'every child goes through a bug period – I never grew out of mine'. Wilson built on ideas like those expressed by Emerson and others to popularise the concept of 'biophilia', the idea that because we humans evolved in nature, we still need nature for a healthy existence – both physical and mental. Experiencing the natural world is not just an agreeable pastime, it's a vital part of our wellbeing.

To truly experience nature, Emerson suggests solitude. He opens the essay *Nature* by suggesting:

To go into solitude, a man needs to retire as much from his chamber as from society. I am not solitary whilst I read and write, though nobody is with me. But if a man would be alone, let him look at the stars.

Inspired by these thoughts, another American nature writer, Henry David Thoreau (1817–62), sought solitude in the woods around his home in Concord, Massachusetts for long periods of time while he worked on his own natural philosophies. Reading the works of the early American environmentalists such as Thoreau and Emerson, along with later pivotal figures like John Muir, there can be no doubt that spending significant time truly experiencing the natural world engenders a gentle and unassuming wisdom. Nor do we have to travel to the wilds of America to connect in this meaningful way. We'll discover comparable inspiration a little further upriver.

Just a few miles upstream from Barnard Castle, close to the village of Cotherstone, I arrive at the confluence of the Tees and the Balder. The River Balder itself is all but invisible, as it emerges from the deep shadows cast by the tunnel of trees that follows the river's lower course. Two narrow roads lead into Baldersdale and the higher reaches of the Balder, although both peter out part-way along the dale. You don't pass through Baldersdale to get to anywhere else, which at least means that the roads along the dale see very little traffic. I've chosen the road that runs to the north of the river since it gives dramatic views of the valley of the Balder. It's May, and although I see almost no cars, my walk is far from quiet. The songs of meadow pipits and skylarks provide a continuous soundtrack from high overhead, while the piccolo calls and songs of oystercatchers drift over stone-walled fields. Over the same fields, the exuberant yips of lapwings accompany their wild courtship aerobatics. On broad butterfly wings, they turn and twist like scraps of paper blown on the wind. At the start of our journey, we watched hypnotic flocks of hundreds of these birds drifting over the saltmarshes of the Tees Estuary, but it's just as hard to tear myself away from this single pair in their remote upland breeding territory.

Eventually, the road offers a view across Hury Reservoir, one of three reservoirs created by damming the Balder. A little further along the dale, a track cuts down towards Blackton Reservoir and across upland haymeadows bright with flowers. These meadows were part of Low Birk Hatt Farm, which was once the home of Hannah Hauxwell. Hannah spent most of her life at Low Birk Hatt Farm, which is where Yorkshire Television producer and director Barry Cockcroft met her in 1972. He'd already produced several documentaries about dalesfolk as part of a Yorkshire Television series called *Country Calendar*, which had proved very popular. So, when a friend mentioned a remarkable woman that he'd met while hiking in Baldersdale, Cockcroft decided to investigate further. Not without some difficulty, he found Hannah Hauxwell, who, by 1972, had spent more than a decade living on her own in a house with no electricity and no running water.

Unlike Emerson and Thoreau, Hannah hadn't sought solitude; it had been thrust upon her when her mother, the last of her family, died in 1961. Since then, she'd run the farm, consisting largely of one milking cow, which produced two calves a year, generating an annual income of just £250–£280. She didn't always relish the solitude, although it's certainly easy to find here, and she was often lonely. However, her quiet life gave her a deep appreciation and love for this remote corner of the dales, together with an enduring wisdom and gentleness that would soon propel her to international fame.

Barry Cockcroft quickly realised that he'd found a truly remarkable person, someone on whom he could base his next documentary. And Hannah was only too happy to have the company when the TV crew arrived, flown in by helicopter with the inevitable mountain of heavy equipment. The resulting film, called *Too Long a Winter*, aired in January 1973. Today (at the time of writing), it's available online and will more than repay the time it takes to watch it. As a TV documentary producer myself, I admire the high production values for what was (when it started out) a local television show. But the film is much more than that, and Hannah was the undoubted star.

Winter came early that year. Heavy snow fell on 12 November, cutting off the dale. In the film, we watch Hannah struggling through the snow, bent double with a haybale on her back as she brings feed in for her cow. Then she goes down to the stream to chisel away thick ice so she can fill her kettle and make tea over a coal burner in a house lit only by oil lamps. Hannah does her washing in Blackton Reservoir, which she always called 'Mississippi', since its expanse reminded her of what she'd heard of that river. She explains in quiet, uncomplaining tones how she has to hang sacks from the ceiling containing what food she can afford, to keep it out of the reach of rats. And she reveals that she treats her animals, in particular her milking cow, as part of her family.

The morning after the film's transmission (on the evening of 30 January 1973), Yorkshire Television's switchboard at its Leeds headquarters was jammed with calls. Shortly afterwards, huge mailbags were dumped at the studio, many containing gifts of all kinds, even postal orders for 50 pence from pensioners who didn't have much more money than Hannah herself. A small army of admirers turned up in Baldersdale, although Low Birk Hatt Farm is so far up this remote dale that no one actually made it that far. Later, a party from Switzerland arrived on her birthday.

Too Long a Winter soon aired nationally, to equal acclaim. Barry Cockcroft became a friend to Hannah and twenty years later, he made a sequel, called *A Winter Too Many*. In the intervening years, continued public donations had allowed Hannah to install electricity at Low Birk Hatt Farm, although, in 1978, a severe winter storm cut off the power and Hannah's false teeth froze solid in the cup by the bed – a reminder of times not long passed when she had to sleep in her greatcoat. After *Too Long a Winter*, Hannah had become something of a celebrity. In the sequel, the crew followed her to London, where she'd been invited to a 'Women of the Year' lunch. She got to stay in a hotel for the first time – the Savoy, no less – but with her quiet and self-reliant daleswoman's attitude, she took the whole thing in her stride.

However, by then, and despite modern conveniences and

more money, Hannah was finding it hard to cope with daily life on the farm. On camera, she speaks of the heart-rending decision to leave Low Birk Hatt. Her cows, companions of many years, go to live with her neighbour. I defy anyone to remain dry-eyed as she drives her animals down the lane to hand them over. She moves to a cottage in the village of Cotherstone, at the foot of the dale and close to the banks of the Tees. Her adventures, however, are far from over.

In 1992, Barry Cockcroft filmed her first venture abroad as she toured around Europe, exploring the famous sites and even meeting with the Pope. Amid genuine sharp curiosity, she still never lost her humility. In an echo of the similar trip by Mark Twain, the series was called *Hannah Hauxwell: Innocent Abroad*. Again, it proved so popular that a second series was commissioned, this time following Hannah around the USA.

When she appeared in *Too Long a Winter*, reviewers referred to Hannah as an old woman, since that was how she looked. Yet, at the time, she was only forty-six; her appearance reflected the hardships she'd endured in the unforgiving landscapes of the dales. It must, however, have been a healthy life. She lived until the age of ninety-one and passed away in 2018. After a service of thanksgiving in Barnard Castle, she was buried in Romaldkirk, not far from the farm where she spent so much of her life. But her legacy lives on.

During her farming years, Hannah couldn't afford to buy expensive pesticides or fertilisers, so, managed in a traditional manner, the hay meadows on the farm grew rank with wildflowers. Those meadows are now in the care of the Durham Wildlife Trust as Hannah's Meadow Nature Reserve, among the finest examples in upland Britain – and they're still a riot of colour in spring and summer. In damper areas, bright-yellow marsh marigolds flower early in the year, followed closely by patches of pink ragged robin. Cuckoo flowers are also open for business in early spring, in time for queen bumblebees to fuel up, before starting to raise their colonies of workers. Orange-tip butterflies join the bees, fluttering around the cuckoo flowers in search of nectar, although the plant serves a dual purpose for this butterfly. Its caterpillars feed on the seed pods.

There are also less-obvious little gems hidden here. Search among the grasses and you may find two small and very strange ferns, moonwort and adder's tongue. Both grow to no more than a few inches and both consist of a sterile leaf, which is concerned purely with the business of photosynthesis, and a fertile leaf, more like a stalk covered in a broccoli-like mass of spore-producing bodies at its tip. You'll also need to keep your nose to the ground to find a frog orchid, one of Britain's least-conspicuous orchids, with a spike densely covered in very unshowy flowers of a pale green or perhaps with hints of purple. Nevertheless, it's always a thrill to find one.

Traditionally managed meadows are rich places to explore – full of flowers and insects, which in turn provide food for birds and mammals. Yet they're now among the rarest habitats in the country. As we've already seen, some 97 per cent have vanished in the years following the Second World War. The all-out drive to intensify farming in the post-war years had less impact in Teesdale than in many other places, since it was protected by its remoteness and by the difficulty of increasing profits from these marginal uplands. Low Birk Hatt wasn't the only farm here to manage its meadows in ways that would have been familiar to many earlier generations of farmers stretching back to the Iron Age. Hannah Hauxwell could remember with startling clarity the joys of haymaking time from her own childhood.

In those days we would always start the haymaking at Hury, a parcel of land we owned two miles down the road ... There were lots of wild roses and foxgloves growing around the lanes and fields and you could smell the hawthorn and rowan tree blossom. Those summer nights at Hury will stay with me forever.[*]

---

[*] Hannah Hauxwell and Barry Cockcroft. *Seasons of My Life: Tales from a Solitary Daleswoman*. Century Hutchinson Ltd. 1989.

Hand in hand with the loss of traditional farming was the abandonment of upland farms as it became harder to make farming up here profitable. In her youth, Hannah recalled a more vibrant Baldersdale, with a large enough population to hold a local agricultural show – the Hury Show – on the banks of the Hury Reservoir.

> Of course, most people have left now – either died or moved away. I'm afraid that's the story of Baldersdale today... the place they abandoned... the dale they left to die... The place was so full once, alive with activity and children. Baldersdale is an empty place now and all that remains for me are memories, sweet memories.

At the end of our journey, we'll explore how repurposed farming subsidies could be used to enhance the conservation value of upland farms like those in Baldersdale, places where nature often hung on in more diversity and abundance than elsewhere, as Hannah remembered from her childhood. But, applied in the right way, new subsidies could also revitalise local communities. For thousands of years, wildlife in Britain has lived alongside farmers. Almost all our landscape is managed in one way or another, three-quarters of it as farmland, and often in the past such management has helped to maintain a much richer natural history than that of today. Yes, we lost our big, charismatic mammals early in farming history, but even so, until the twentieth century, a vibrant natural world still existed on farmland and in managed woodlands, where humans replicated some of the ecological effects of the big grazers in the wildwood.

The sunlit floor of coppiced woodlands abounds in spring flowers and butterflies, and hay meadows grow tall with flowers from spring through summer and, before cutting in late summer, hum with bees and other insects. Hannah's Meadow shows what can be achieved with appropriate management, and appropriate subsidies. As it becomes ever harder for hill farms to turn a profit from conventional farming, it surely makes sense for farming subsidies to be focused on supporting a combination of upland rewilding and extensive rather than intensive farming. This would recreate some of the natural abundance found on farming landscapes of the past as well as creating a source of income from eco-tourism. From Hannah's Meadow, I can see a great vista lying beyond Blackton Reservoir, fields and moors stretching away into the haze of a warm spring afternoon, land that with a new mindset could be revitalised for all our benefit.

Hannah Hauxwell was an extraordinary person, but more extraordinary still was the effect she had on people across the world. Clearly, Hannah struck a chord with all those who saw the films. In their generous donations of gifts, I'm sure in part they simply wanted to help someone who was living a hard and deprived life in the heart of England at the start of the 1970s – a heartening demonstration of an underlying generosity in human nature. But the response was so massive that I think she touched an even deeper part of their humanity. Her tranquil acceptance of a life without many material possessions, compensated for by the view from her back door and by the landscapes and creatures of the dales, was perhaps a reminder of where enduring happiness really comes from – not from the ostentatious trinkets of conspicuous consumption, such as those housed in Bowes Museum, but from close contact with a more natural world. From Emerson in Barnard Castle to Hauxwell in Baldersdale, it's a truth which has often been expressed and rarely acted on.

9

Gifts of the Earth

Middleton-in-Teesdale to High Force –
Mines, Mires and Meat-Eating Plants

> Sweet vale, though we sing in thy praise,
> Yet life has its stern earnest duties;
> Thy sons in all countries are found,
> Poor men cannot live by thy beauties.
>
> RICHARD WATSON,
> 'Lovely Sweet Vale of the Tees', 1884

F ollowing the river upstream, our journey has taken us through ever-more-rural settings, across farming landscapes of pastures and crops. It might seem that we've finally left behind the industrial landscapes of the lower river but, as we approach Middleton-in-Teesdale, that turns out not to be the case. In the nineteenth century, as the lower river was being encircled by iron and chemical industries and an ever-increasing human population, so the dales of the upper river were suffering their own Industrial Revolution – although instead of a steel river, here it was a river of lead.

The North Pennines have been mined for lead for millennia but by the middle of the nineteenth century, Middleton-in-Teesdale had become the heart of one of the country's largest lead-mining operations. Not that arriving in Middleton-in-Teesdale today feels like walking into a scene of industrial devastation. Its wide main street is bordered by a long village green and flanked by attractive stone or whitewashed houses. It's a tranquil spot where hikers mingle with locals and where the conversations overheard in the teashop are mostly about sheep. However, the legacy of a very different past is still apparent if you know where to look.

I've always liked Middleton-in-Teesdale. For me, this quiet town is the gateway to Upper Teesdale – a portal to a wilder world of fells and dales. Moreover, I have a deep connection to this place; one branch of my family hails from this town. Sometime in the middle of the nineteenth century, my third great-grandmother, Mary Brunskill, migrated downriver from this area and, presumably either unmarried or widowed, gave birth to a son, Isaac Brunskill, near Stockton-on-Tees. Isaac later found work in the expanding iron industry in Middlesbrough before ending his life as a pauper in the Middlesbrough workhouse. Further back still, in the early 1800s, my fifth great-grandfather, Robert Brunskill, born and bred in Middleton-in-Teesdale, was a lead miner – and a tiny part of a growing industry that would soon turn

Middleton-in-Teesdale into an industrial boomtown. Between 1801 and 1871 the population of the town quadrupled, and 90 per cent of these people worked in lead-mining and processing.

This remarkable transformation was made possible by events that happened in an unimaginably distant past. Turn back time a few hundred million years and instead of standing on a high fell, you would be standing by the shores of a tropical ocean. This period of deep history, the Carboniferous, which lasted from about 350 to 300 million years ago, built the foundations of England's North-East, both physically and economically. The world was a very different place back then. Continents and continental fragments were sliding over the surface of the globe – as they've always done and still do today. However, at this time, they were involved in history's biggest pileup, albeit in extreme slow motion. All the landmasses on the planet were colliding with each other, slowly building a single supercontinent known as Pangaea.

All this activity caused great changes in the region that would one day be called the North Pennines. Sometimes the whole area lay under a warm ocean, cloudy with tiny marine organisms, many of which built shells from calcium carbonate – chalk. They lived and died in such numbers that their chalky remains built up into thick layers which would later be compressed and transformed into the hard Carboniferous limestone that now forms the backbone of England.

As sediments built up, shallow seas gave way to swampy lowlands. Around this time, life invented trees, although nothing like those of today's forests. These ancient trees were giant horsetails, club mosses and ferns, and they thrived in the warm Carboniferous swamps. These alien forests were only possible because plants began making liberal use of a substance called lignin – a chemical that can crosslink the fibres of cellulose that run through the walls of plant cells, turning them into one of the toughest of all biological structures – wood. It was so strong that it could support the massive weight of giant plants – as it still does today.

Lignin does its job by being almost indestructible. It's tough work for fungi to break lignin down (consider how long a fallen tree lasts), but when it first evolved, fungi had no answer to this stuff. Dead trees lay where they fell, to be slowly buried and, over time, to be compressed into coal. When the seas returned, more limestone was laid down, followed by another cycle during which coal forests predominated. Jump forward 300 million years and the Carboniferous bedrock of the North-East provided limestone for the blast furnaces along the lower Tees and created the extensive coalfields of Durham and Yorkshire that fuelled the furnaces. But geology wasn't finished with the North-East just yet.

Towards the end of the Carboniferous Period and at the very beginning of the subsequent Permian Period, a 'small' supercontinent called Laurussia ground into the much bigger landmass of Gondwana as Pangaea began to take shape. The collision distorted and fractured the rocks that would later become the North Pennines, creating swarms of faults. The upheaval also reactivated older faults. Shortly after this, continental movements stretched the crust along a north–south axis, widening all the faults and allowing a great belch of magma to well up from deep within the Earth. The magma rose along vertical faults, called 'dykes', then, wherever it could, it flowed in great horizontal sheets between existing rock layers.

It solidified into a hard, erosion-resistant rock called quartz dolerite by geologists and whinstone by local miners. Miners also referred to the great horizontal slabs of whinstone as 'sills', so this whole igneous complex has become known as the Whin Sill. Incidentally, sill has now entered the geologist's lexicon as a name for any similar horizontal igneous rock formation, but the Whin Sill was the first – and it's responsible for many of the North-East's most iconic landscape features.

Whinstone is so hard that it erodes more slowly than the

surrounding rocks and therefore often stands proud of these softer rocks as the long crags or ridges of hills that are so characteristic of the landscapes of Durham and Northumberland. From Teesdale, the Whin Sill runs south through Luredale. In the opposite direction, it runs north through Weardale and in Northumberland creates the impassable cliffs that form natural ramparts along sections of Hadrian's Wall. Further north still, whinstone is responsible for the lines of rugged, crag-scarred hills that make up the wild landscapes of the Wannies.

On the coast of Northumberland, the Whin Sill outcrops again to create the steep hills which serve as foundations for both Dunstanburgh and Bamburgh Castles. Nearby, Lindisfarne Castle is perched atop Beblowe Crag, another outcrop of whinstone that gives Holy Island its characteristic profile. Off the coast, further outcrops of the Whin Sill have resisted erosion by the waves to form the Farne Islands, fifteen to twenty of them, depending on the state of the tide.

It's been estimated that some fifty cubic miles of magma welled up to create the Whin Sill and that this flowed outwards from the vertical dykes eventually to cover nearly 2,000 square miles. So much hot magma flowing through the surrounding bedrock had far-reaching effects. Where the magma ran close to shale rocks, the shales were converted into a much harder rock, called whetstone by the Teesdale miners, although I don't know if these rocks were ever used to actually sharpen blades. When magma intruded into the Carboniferous coal measures, its heat altered the main constituent of coal, the organic equivalent of a mineral called vitrinite, which is composed of cellulose and lignin, the last earthly remains of those vast swamp forests.

In hotter zones, closer to the magma, vitrinite becomes blacker and shinier. The degree of its shininess – or, more properly, its reflectance – can be measured accurately, and the distribution of these different types of coal across the region helps in understanding how the Whin Sill formed. Some types of vitrinite are best explained by repeated heating and cooling and suggest that the Whin Sill was created by several distinct episodes of intrusion. Of more practical interest, the shinier and blacker the vitrinite, the better the coal. So, the way in which the Whin Sill was intruded shaped the patterns of high-quality coal in the Durham coalfield, which in turn played a role in the birth of railways as a means of transporting that coal to Stockton, and in the origins of the town of Middlesbrough as a deepwater port to export the coal. Much of the recent history in the first part of our journey along the Tees was shaped by events in an unimaginably distant past. Yet, the Whin Sill also has a big part to play in the final stages of our journey.

Where it flowed through the Carboniferous limestone that makes up much of the Pennines, it cooked these rocks and transformed them into a much softer and more crumbly kind of rock known as sugar limestone. Upstream of Middleton-in-Teesdale, sugar limestone creates some very special environments on the high fells, which are home to an assortment of intriguing plants, together known as the Teesdale Assemblage. We won't reach these fells until the last stage of our journey, so we'll save our exploration of the sugar limestone landscapes until the next chapter. Of more importance to the industrial history of the North Pennines, as the Whin Sill magma cooled, it contracted, which created a maze of open channels through the rock. Mineral-laden water percolated through these channels and, warmed by molten magma deep below the surface, the dissolved minerals were precipitated to produce veins of different mineral ores.

One of these ores is lead sulphide, more widely known as galena, the most common ore from which the metal lead can be extracted. Some veins were only inches thick; others were many yards wide and well worth exploiting. The Whin Sill was so vast that the North Pennine lead field – comprising Teesdale, Weardale, South Tynedale and the Derwent Valley – became the most important lead-producing area in Britain.

Lead-mining has a long history in the North Pennines. The Romans extracted lead here, right on the northern frontier of the Empire. In the early Middle Ages, Bede (672/3–735), a monk based in the North-East who left valuable accounts of life in Anglo-Saxon England, recorded that lead miners from Swaledale, to the south of Teesdale, had to haggle with merchants in Catterick to obtain fair prices for the fruits of their labours. The prince bishops of Durham exploited the Durham dales for lead (along with silver), and in the sixteenth century, the Bowes family, whom we met in the previous chapter, were busy digging lead ore from extensive mines around their family seat at Streatlam Castle.

At the beginning of the sixteenth century, lead-mining and smelting were growing in importance, initially spurred on by King Henry VIII's desire to make England independent of Europe (funny how little such things have changed). By 1600, Britain was one of the largest lead producers in the world. Lead was also produced from centres on the Mendip Hills in Somerset and the Derbyshire Dales, but it was the ore fields of the North Pennines that attracted the attention of the London Lead Company.

The London Lead Company had been buying leases in Teesdale, including some from the Bowes family, since the middle of the eighteenth century, but in 1815, it established its northern headquarters in Middleton-in-Teesdale. The superintendent lived in grand style at Middleton House (which also housed the company's offices), but the company also provided for its workers. The London Lead Company was owned by Quakers, whose beliefs embodied a far more benign attitude to their workers than was common in the period. The owners of the London Lead Company saw a need to provide better living standards for their workers, so, in 1824, New Town, on the south-eastern side of the village of Middleton-in-Teesdale, began to take shape to provide homes for the growing workforce.

Even so, the houses reflected the status of the workers. Surveyors, for example, might expect one of the grander dwellings on Masterman Place, whereas miners had to content themselves with a one-up, one-down row house, such as those on Ten Row, now renamed River Terrace. These houses, though, did have long gardens for growing vegetables and some even came with a pig sty (pig not included). The company also built schools and a workers' library, but it expected its employees to share the Quaker philosophy of temperance and avoid the evil drink. New Town is still entered through an impressive stone archway but in the nineteenth century the arch housed gates that were locked shut at night to encourage sobriety. I can't imagine that went down at all well among labouring men.

Despite the enlightened attitude of the Quakers, the life of a Teesdale miner was hard beyond imagining and it began at an early age. Young children worked near the entrance to a mine, where they crushed the lead ore into small fragments with a hammer – all for the meagre wage of fourpence a day.

> Me father were a miner, he lived down in t' town,
> 'Twere hard work and poverty it always kept him down;
> He aimed for me to go to school, but brass he couldn't pay,
> So I had to go washing rake for fourpence a day.

These lines are taken from the song 'Fourpence a Day', which describes the life of a child labourer in Middleton-in-Teesdale in the latter half of the nineteenth century. The song was made more widely known in the 1950s by Ewan MacColl, a singer-songwriter whose work also gave a voice to modern miners. For his seventieth birthday in 1985, Arthur Scargill presented MacColl with a miner's lamp in appreciation for the songs that he'd written in support of the miners' strikes during the 1970s. MacColl was also one of Britain's most prolific collectors of old folk songs. Over his life, he rescued a rich heritage of material describing the lives of ordinary working people in their own words. To listen to these songs is to get a sense of the harsh lives of miners and labourers, steelworkers and fishermen, if not at first hand, then far more vividly than any book or museum exhibit can ever portray.

MacColl learned 'Fourpence a Day' from the singing of John Gowland, a retired miner from Middleton-in-Teesdale, and he committed the lyrics to print in the early 1950s. Many of these songs had never been written down and were disappearing fast through the middle years of the twentieth century. Life in Britain was transformed in these post-war years – a time when younger generations saw no value in learning the old songs. Yet, in the right hands, these songs still have wide appeal. MacColl collected many songs from Teesdale, including one that is now widely known right around the world.

In 1947, MacColl visited Mark Anderson, another retired miner in Middleton-in-Teesdale, who sang him a song he called 'Scarborough Fair'. MacColl published this in a book of Teesdale songs, from where another well-known English folk singer, Martin Carthy, learned it. Carthy then taught it to an up-and-coming American folk singer called Paul Simon, who'd been working as a duo with another singer called Art Garfunkel since they'd been at school together. In 1966, the pair recorded 'Scarborough Fair' on their seminal album *Parsley, Sage, Rosemary and Thyme*, named from a line in the chorus of the song. Simon and Garfunkel would soon become household names across the world and once sang 'Scarborough Fair' for an audience of 50,000 crowded into Central Park in New York City. It's an awfully long way from a living room in the house of a retired Middleton-in-Teesdale lead miner to Central Park, but such is the power of these songs.

The road through town eventually makes a sharp bend as it crosses Hudeshope Beck, which joins the Tees here. Just before this, Town Head forks off from the main drag and brings me to the church of St Mary the Virgin,* where I'll find another miner who has left a legacy of evocative descriptions from the height of the lead boom. One of the stained-glass windows portrays a bearded man sitting next to a narrow railway line leading into the blackness of a mine. The man's name was Richard Watson (1833–91), and he's often called the Teesdale poet, or, as inscribed on his headstone in a quiet corner of the churchyard, 'The Bard of Teesdale'. His view from the churchyard embraces the valley of the Tees upstream from Middleton-in-Teesdale, where the river flows past craggy cliffs of whinstone – and where he spent his working life.

The son of a lead miner, he was born in one of the tiny houses on Ten Row (River Terrace). He went to the company school but had to leave aged only ten since his father had become seriously ill and was no longer able to work. The ten-year-old Watson now had to earn a living for his family. He could have been the voice in the song 'Fourpence a Day', as he began work, washing and grinding lead ore – although in his case for the princely sum of sixpence a day. The mines were often so distant from the town that the workers, children included, hiked up on Monday, stayed in camps or 'lodge shops' at the mine head, sleeping three or four to a verminous and stinking bed, before hiking back at the end of Friday.

Watson began writing to entertain his fellow miners but his work became more widely known after a local newspaper, the *Teesdale Mercury*, began to publish his poems. He was often asked to give recitals at local events, although he never made much money and died in poverty at the age of fifty-eight – the lot of a great many miners of lead, coal or iron.

In one poem, called 'My Journey to Work', Watson describes the seven-mile hike from his home at Holwick, south of the

---

* At the time of writing, the church is closed owing to extensive dry rot, although the churchyard is still open.

Tees, to a mine at Eggleshope, north of Middleton-in-Teesdale. There, he stayed in the lodge shop until his return trek on Friday. In this, and many of his other works, he contrasts the natural beauty that he sees all around him with the hard life of a miner, although he also sees the paradox that it's the mines that provide a living, however meagre, to a great many people here. It's the same conflict I still feel when I see the economic distress of so many people in my hometown of Middlesbrough, after the demise of heavy industries, weighed against the urgent need to rebuild the natural ecosystems of the lower Tees – and indeed those across the whole of Britain.

As Watson began his weekly trek to the mine early each Monday morning, he heard and saw only the natural world of Teesdale, not the hard week in the dark mine that lay ahead of him.

'Tis Monday morning and Apollo bright
Gilds the far eastern hills with golden light;
To hail the rising orb of day, I hear
The cheerful rousing voice of chanticleer;
From the whin cliffs the wild rock pigeons fly,
And starlings there keep up chattering cry
All animated things seem blythe and gay
Pleased with the prospect of a pleasant day.

Parts of Watson's weekly walk are now a heritage trail, taking in the same sites that inspired his words. I'll be walking sections of this as I follow the river upstream from Middleton-in-Teesdale, so we'll be hearing more of this remarkable miner-poet a little later in our journey.

By the 1890s, the richest veins of lead in the North Pennines were worked out. In any case, competition from Spanish lead had already made working the Teesdale mines uneconomical. In addition, other metals were growing in importance, not least iron from the growing number of furnaces scattered along the banks of the lower Tees. The London Lead Company ceased operations here in 1905 and many of the miners were forced to seek work elsewhere as Middleton-in-Teesdale slowly returned to being a quiet country town.

To rejoin the Teesdale Way, which is now coincident with the Pennine Way running along the south bank of the river, I need to head south, out of Middleton-in-Teesdale, along Bridge Street. My walk takes me past an ornate fountain, built in 1877 to honour Robert Walton Bainbridge on his retirement as superintendent of the London Lead Company – a lasting reminder of the days of lead-mining in Teesdale. However, there are other legacies of the mining past etched into the landscapes of Teesdale, and I'll find these on my journey ahead. So, in leaving Middleton-in-Teesdale, we aren't leaving the story of lead-mining behind.

Bridge Street leads me across Middleton Bridge, over the Tees and past large sheds which house a cattle market, mercifully quiet today. Turning right onto a rutted track, I make my way through farmland and across Crossthwaite Beck. The river briefly meanders away from the path, then back again for a few paces, before once more disappearing from view. After a mile or two through somewhat uninteresting farmland with no river to keep me company, the path brings me close to Park End Wood, a tiny patch of downy birch, ash and hazel growing on a small hill of whinstone. From here, the Teesdale Way rejoins the river and winds its way along the bank as the river cascades over stones and around boulders. As I approach the river, a dipper takes off and flies a short distance upstream. Its dark body with its bright-white breast is soon lost in the broken water tumbling over the stony bed. It flushes again each time I approach, until it reaches the end of its river territory, when it quickly doubles back and once again is lost on the turbulent river.

The noise of the river and the hypnotic movement of the water are wonderfully meditative, and I'm surprised to find myself already approaching Scorberry Bridge, a footbridge linking the south bank with the village of Newbiggin to the north of the river. I'm going to make a detour from the Teesdale Way here to join part of the route used by Richard Watson and to visit Newbiggin, where he often stopped to rest and catch up on local gossip.

Newbiggin's reached, where miners often stop
To light their pipes at Willie Gibson's shop.
A blacksmith Willie is, of well-tried skill
Whom we find hard at work when we will;
When'er we meet he greets me with a smile,
Enquires the news, and bids me rest a while.

From Newbiggin, Watson continued on to Coldberry. Here, on the ridge above me, there's a distinctive 'V'-shaped notch on the horizon, a neat slice cut from the hill – Coldberry Gutter. Up close, Coldberry Gutter is a long, steep-sided valley that may owe its origins to the lead industry. It's thought to be a 'hush' – a relic of an early form of hydraulic mining called 'hushing' which was practised widely in the eighteenth century. There are many hushes scattered across the dales, although Coldberry Gutter is the largest, gouged for over a mile through the hillside. Hushes were created by harnessing the power of flowing water. Miners built dams across streams above areas they thought might contain lead ore and, once a large enough head of water built up, the dam was breached. The gushing water stripped away peat and soil to expose any buried lead or silver ore, and in the process often carved a small 'V'-shaped valley.

Hydraulic mining is one of the most destructive ways of extracting ore. In Teesdale, eighteenth-century technology and engineering limited its impact to a scattering of man-made valleys that are now part of the scenery and archaeological curiosities. By the nineteenth century, new technology allowed mines to burrow beneath the dales in search of ore, but else-where nineteenth-century engineering skills unleashed the full destructive power of hydraulic mining. On several occasions, I've walked over barren moonscapes of bare mud on the flanks of the Sierra Nevada Mountains in California, the legacy of the California gold rush. Earlier on our walk, we saw that iron ore was discovered in the Cleveland Hills at almost the same time that James Sutter picked up a nugget of gold exposed in a streambed in the Sierra Nevada Mountains, and in 1849, the remote and still-pristine Sierra Nevada filled with thousands of people hoping to get rich quick.

Gold was simply lying in the streambeds, gently eroded out from deposits in the hillsides over thousands of years. However, these easy pickings were soon exhausted, and the 'forty-niners' couldn't wait for the slow pace of natural erosion to reveal more. They dammed streams high in the mountains and built thousands of miles of wooden sluices (which denuded huge tracts of forest) to channel the water to those hillsides, which might be hiding more gold. The sluices delivered water to specially constructed nozzles under such enormous pressure that it was said that a man couldn't swing a crowbar through the water gushing out. The water exploded from the nozzle at 150 miles per hour, powerful enough to kill a man standing more than 50 yards away and more than enough to blast apart whole hillsides.

The rivers ran so thick with mud that they could hardly run at all. By some estimates, the worst-affected rivers won't be entirely free of mud until the year 3000. Mercury was used to extract trace amounts of gold from the flushings, so the accumulating mud was as toxic as anything from the worst days of pollution on the Tees Estuary, and still today little grows there. Luckily, Teesdale never experienced this level of destruction, at least not at the hands of the lead industry.

Some geologists and archaeologists now think that while the smaller hushes are the results of the activity of lead miners, Coldberry Gutter might have natural origins – carved by water from an ice-dammed lake which formed during the final stages of the last Ice Age. As the climate warmed, the ice dam breached, releasing far more water than any lead-mining dam could ever hold. As impressive as Coldberry Gutter is, though, it's tiny in comparison to some of the features carved in similar fashion elsewhere, by the sudden release of meltwater at the end of the Ice Age.

In the late stages of the Ice Age, beginning some 15,000 years ago, ice dams created truly enormous lakes of meltwater, followed by floods of unimaginable scale when these dams gave way. Almost overnight, in geological terms, one such flood in Greenland scoured multiple canyons that in some cases reached twice the length of Arizona's Grand Canyon. Today, these canyons are hidden beneath the ice, more than a mile thick, that forms the Greenland Ice Sheet, but if they were ever revealed, they would be one of the most dramatic landscapes on the planet. In western North America, the evocatively named Channeled Scablands of Washington State also owe their origin to an immense flood, this time unleashed by the breaching of ice dams holding back the waters of Glacial Lake Missoula (which covered large parts of western Montana to a depth of nearly 2,000 feet).

At its peak, this flood carried the equivalent of eighty-five Amazon Rivers – and this happened not once, but perhaps a hundred times as the ice retreated at the end of the Ice Age. The landscape is stripped back to the skeleton of bedrock, crossed by steep-sided canyons carved by the water. A little closer to home, the Strait of Dover was also gouged out of the chalk hills that once connected Britain to France by a massive outwash flood from a lake in the area now occupied by the North Sea. If Coldberry Gutter is an outwash 'canyon', then the impact of outwash floods on Teesdale has been far more modest than elsewhere.

However, the scenery here bears other marks of lead-mining beyond the tiny valleys of the hushes. Travel through Coldberry Gutter, and to the east lies Coldberry Mine, the entrance sealed with a metal grill and surrounded by derelict buildings. Beyond these are hills built from the spoil of mining. The soil on these hills has high concentrations of metal ions, toxic to most kinds of plants. Only a few specialised species survive here, in what ecologists call Calaminarian grassland. Such grasslands do occur naturally, for example, on mineral-rich serpentine rocks or where mineral veins outcrop at the surface, but most examples are on spoil heaps, the relics of mining right across the country. Far from being industrial wastelands, the grasslands are of great interest to ecologists, partly because the toxicity of the metal ions prevents the succession of grassland into other habitats. Left to its own devices, and in the absence of heavy grazing, most grassland soon becomes covered in scrub and eventually trees. Calaminarian grassland, on the other hand, remains as grassland and is therefore home to some of our rarest – if often most obscure – grassland plants.

Spring sandwort is a typical plant of these grasslands. It's tiny and low-growing, with thin leaves easily lost among those of grasses. However, in spring, it produces a mass of starry-white flowers that would grace any alpine bed or trough. More conspicuous are the heavy-metal-adapted races of thrift and sea campion, both common plants around Britain's coasts. Sea campion produces a succession of large, bell-shaped flowers that flare into a skirt of brilliant-white petals. In contrast, the clumps of thrift are covered in pink pom-poms.

Teesdale's Calaminarian grassland is also one of the best places in Britain to find a much rarer plant – alpine penny cress. As its name suggests, it's more common in the Alps, while in Britain, almost all of the very few places where it can be found are natural or post-industrial grasslands that are laced with the ions of heavy metals. Alpine penny cress grows as a tiny rosette of leaves, easily missed if not in flower, but in season it produces an outsize head of white flowers that seems far too big for the tiny plant beneath.

Retracing my steps through Coldberry Gutter and back through Newbiggin, I rejoin the path that Richard Watson trod each week. I'm heading upstream, in the direction that would have taken Watson home at the end of a back-breaking week, yet along this stretch his senses were still attuned to a more natural world in which he sought respite from the hard life of working men.

The ears regaled with many a pleasing song
Of happy birds that flit from tree to tree
And seem, unlike our race, from sadness free.

Perhaps one of those birds was the 'Teesdale canary', more widely known as the siskin, the males of which more than deserve their comparison to canaries. On my way back to the river, I arrive at the visitor centre at Bowlees, ready to refuel on some of the fine cakes on offer – but before I can indulge myself, I'm distracted by the bird feeders hanging outside. They're alive with a twittering flock of siskins. The males all sport dapper black caps and have bright-yellow wing bars and cheeks, and yellow breasts. The females are just as striking in their own way. More subdued in colour, they, too, have bright wing bars, although they are otherwise covered in soft streaks. They're all hanging off the feeders at improbable angles, demonstrating an agility normally used to extract the seeds of pines, birches and alders, which make up much of their natural diet.

I refresh myself with cake and coffee while watching the antics of the siskins, along with blue tits, which, although much smaller than the siskins, are more than feisty enough to claim a place at the feast. From the visitor centre, there's one more detour I want to make before finally returning to the river – a short trek up Bowlees Beck, a tributary of the Tees, as far as Summerhill Force and Gibson's Cave. As usual, I've hardly started this short walk before I'm distracted, this time by what looks like a small quarry, long abandoned and with a damp, mossy floor. On the higher, grass-covered hummocks, there are lots of flower spikes of common twayblade orchids. If you envisage orchid flowers as exotic, flamboyant and colourful, then, in the case of the common twayblade, think again.

Twayblade flowers are tiny and green. The whole plant, however, is the embodiment of architectural elegance. Two broad, elliptical leaves (the 'tway blades') are held a few inches above the ground, in perfect symmetry, and from the centre of the leaves, a long flower spike carries dozens of small green flowers which all glow with an ethereal vibrance when the sun catches them.

On the damper quarry floor, I notice a scattering of large bluish-purple flowers that, from a distance, might be mistaken for violets. However, a closer inspection reveals that the flowers are held two or three inches above rosettes of pale-green leaves, each of which looks remarkably like a stranded starfish. They're common butterworts, and those rosettes of leaves are deadly traps for any small insect. Butterworts are one of the many kinds of plants that have adopted the very un-plant-like habit of eating animals. Carnivory has evolved on many separate occasions in more than a dozen unrelated families of plants, usually in response to a similar ecological pressure – a lack of nutrients in the soil. Many are plants of waterlogged ground in bogs where inorganic nutrients, such as nitrates, are in short supply. On the other hand, there's plenty of nitrogen in the air above the bog, in the form of countless insects buzzing around. All a plant has to do is catch and then digest some of these insects and it will have all the nitrogen it needs.

Plants have evolved many ingenious ways of doing this. In the same family as butterworts (Lentibulariaceae), the bladderworts grow tiny suction traps. Half a dozen types of bladderworts grow in Britain in acidic ponds and lakes, although elsewhere, other species also grow in wet soil. Scattered among their leaves are the tiny bladders that give these plants their name. The bladders are hollow structures with a small entrance that is normally tightly sealed. The plant pumps water out of the bladder, which creates a strongly negative pressure inside. The mouth of the bladder is fringed with tiny hairs that act as triggers and when a small aquatic bug brushes against one of these, the bladder cracks open and, since nature abhors a vacuum, water rushes in with such force that the tiny creature is swept inside – and to its doom.

Butterworts are not quite so spectacular. Just like the more familiar sundews, they rely on sticky leaves. The upper surface of each leaf is covered with tiny glands that secrete viscous mucus. A small insect landing on the glutinous leaf soon finds itself gummed up, but its struggles to escape only stimulate the plant to release more mucus around the insect. Once its prey is subdued, the plant then releases digestive enzymes from a second type of gland scattered over the leaf surface. Sometimes the whole leaf rolls up along its length to bind its prey more firmly and to retain the nutritious insect broth that accumulates as digestion proceeds. Eventually this liquid fertiliser will be absorbed by the leaf. All that remains of the insect is its tough outer cuticle, and a close-up view of a butterwort leaf often reveals a veritable graveyard of the plant's previous victims.

I have a particular fascination for carnivorous plants. Over the years, I've tracked down giant rajah pitcher plants in the rainforests of Borneo and waded across bogs in North America through spectacular displays of three-foot-high white-topped pitchers – yet wherever I find these plants, I find myself entranced by their intriguing adaptations and extraordinary lifestyles. Even so, time is pressing. It's only a few minutes further upstream along Bowlees Beck to Gibson's Cave and Summerhill Force.

Gibson's cave is hardly a cave at all – more a rock shelter behind Summerhill Force, a small waterfall tumbling over a ledge of limestone. The water has carved out a recess in the limestone that, according to local legend, served as a hideout for a sixteenth-century outlaw called William Gibson, who was on the run from the authorities at Barnard Castle. Since he was something of a local hero, people brought him food and clothing while he remained hidden behind the curtain of foam of Summerhill Force.

Today, this hidden lair is serving as a hunting ground for a pair of elegant grey wagtails. These birds are common along the upper Tees and along its smaller tributaries, and often share their river territories with dippers. Neither bird ever sits still – but whereas dippers dip, wagtails wag. Grey wagtails have tails almost as long as the rest of their bodies, which they wag up and down continuously as they pick their way over stones in search of their insect food. Dippers, with their short, stumpy tails, simply bob their whole bodies up and down, but both birds seem imbued with the ceaseless energy of the fast-flowing streams and rivers on which they live.

Incidentally, as we are about to find out, 'force' is a common name for waterfalls in this part of the world. It betrays the heavy Viking influence along Teesdale, since it derives from the Old Norse word *foss*, for waterfall. 'Force' is commonly used in other Viking areas, such as the neighbouring Lake District. Further north, in Weardale and across Scotland, waterfalls are usually called 'linns', a name of Anglo-Celtic origins for a waterfall or for the pool at the base of a waterfall.

Summerhill Force is an unusual waterfall in that it cascades over a lip of limestone, which is soft enough to have allowed the erosion of Gibson's Cave. The larger waterfalls of the dales are created where water flows over outcrops of the much harder Whin Sill. Walking from Bowlees back to the Tees brings me to one such waterfall – Low Force. Not so much a waterfall, it's a series of small cascades created by a succession of whinstone ledges. Low Force empties into a narrow gorge a little further downstream, which is spanned by a precarious footbridge, the Wynch Bridge. This is the only way over to the south bank to rejoin the Teesdale Way – across swirling, Guinness-dark, peat-stained water. Wynch Bridge is a suspension bridge that swings perilously over the gushing water – and it stands on the site of a much older bridge that proved a little too perilous.

The first bridge was built in 1704, when, according to the 1848 *Journey through Teesdale*, written by one Francis P. Cockshott, it may have been Europe's first ever suspension bridge. It was later replaced by another bridge, built in about 1741, which was no less unstable. The antiquarian William Hutchinson described the somewhat hazardous crossing.

The bridge is 70 feet in length, and little more than two feet broad, with a handrail on one side, and planked in such a manner that the traveller experiences all the tremulous motion of the chain, and sees himself suspended over a roaring gulph, on an agitated, restless gangway, to which few strangers dare trust themselves.

I must say, I shared his feelings crossing the bridge that stands there today – but perhaps Hutchinson's trepidation was more deserved. His bridge collapsed in 1802 as eleven people were crossing. Two fell into the roiling water and although one was rescued, the other died. Again, the bridge was replaced, this time with the one that stands there today.

These bridges served as vital links for miners like Richard Watson, who lived in Holwick to the south of the river but had to reach mines situated to the north, and crossing 'the creaking chain bridge' twice a week was part of 'My Journey to Work'.

A well known path from Holwick to Bowlees,
Where Winch Bridge spans the verdant banks of Tees.
When to the roaring river drawing near,
Its rumbling sound strikes loudly on the ear, –
Foaming and dashing in its rapid course,
O'er the rough grey whin rock named Little Force

Low Force is sometimes called Salmon Leap because, as powerful as the flow through the channels is, salmon are able leap the cascades and falls to reach gravel spawning beds upstream. The fish that we left struggling to cross the Tees Barrage at Stockton have followed our journey every step of the way. Some spawn in the stretches of the Tees or its tributaries that we've already explored, including Bowlees Beck and the Rivers Lune and Balder, although the more adventurous push on upstream beyond Low Force.

The extraordinary power of salmon as they leap waterfalls must be seen to be really appreciated. I've admired their stamina on several rivers around Britain, although one of the most spectacular locations I've visited is the Falls of Shin in northern Scotland. Here, a viewing platform perched on the edge of a rocky gorge gives superb views of the falls and of spectacular leaping salmon. The bodies of these fish are the epitome of streamlining and seem to be all muscle as they make headway against seemingly overwhelming currents. A prodigious leap from the pool at the base of the falls propels a salmon halfway up the cascade. From there, it surfs through foaming whitewater with powerful thrusts of its tail until it reaches slack water at the head of the falls, where it promptly disappears, no doubt to catch its breath – this is, after all, literally a breathtaking experience, for both fish and fish-watcher. I've never seen salmon leaping falls on the Tees, but plenty must put on performances as impressive as the salmon at the Falls of Shin, as they swim upstream to find gravel beds in which to spawn.

Some of the best spawning beds are in Baldersdale, home to Hannah's Meadow. Unfortunately, even though the salmon have struggled this far upriver, they still face problems. In the past, both the practice of hushing and the washing of lead ore at the entrance to mines flushed copious amounts of toxic heavy metals into the river and poisoned the gravel spawning beds. By the mid-nineteenth century, according to the Commissioners for British Fisheries in 1861, the upper Tees was becoming as polluted as the lower reaches, where the proliferating industries were beginning to have major impacts. In more recent times, the construction of reservoirs in the Pennines to supply water to the industries at the river-mouth altered the flow of the rivers, which in turn affected the supply of the right grade of gravel. This is critical to the survival of eggs and fry.

We saw earlier that a lot of effort and money has been expended on restoring salmon and sea trout to the Tees, but all this is pointless without the right conditions in which they can spawn future generations. One way of improving a river for spawning salmon is to create new spawning beds and, to that end, thousands of tons of gravel have been dumped in the River Balder. Similar measures have been taken on other rivers as far away as Devon and Somerset, but the real solution is to remove unnecessary barriers to the upstream journeys of not just salmon and sea trout, but of all migratory fish. We don't have a detailed inventory of all the structures blocking rivers, but it's widely thought that there may be 1.2 million dams, weirs and other barriers impeding rivers across Europe, of which 100,000 are obsolete and serve no purpose other than making life difficult for migratory fish. Across British rivers, there are some 50,000–60,000 obstructions, of which perhaps 10 per cent are redundant. Worldwide, freshwater life is declining, impacted by pollution, water extraction and barriers across rivers and streams of all sizes. Overall, there's been an 84-per-cent fall in abundance among freshwater creatures, but migratory fish have declined by an astounding 96 per cent. The obvious solution is to get rid of any unnecessary barriers, and in many places, this is exactly what's happening.

On once-prolific salmon rivers along America's Pacific Coast, some very substantial dams have already been removed so that ancient spawning grounds are now accessible again. A large dam has been removed from the Sélune River in northern France (the biggest so far), but substantial dams are also being removed from the Hiitolanjoki River in Finland, giving salmon access to spawning beds for the first time in a century. The EU has set a target to create more than 15,000 miles of free-flowing rivers over the next decade. In Britain, however, the aspirations are considerably less ambitious. Even many of the fishways around barriers are not fit for purpose, as we discovered back on the Tees Barrage. Yet, dam removal rejuvenates a river with remarkable speed. After the removal of a tiny weir on Greatham Creek, which flows into the Tees Estuary, eels responded quickly by moving upstream in much larger numbers.

I, too, need to move upstream. The Teesdale Way continues to follow the south bank of the river above Low Force, where the Whin Sill, in resisting erosion by the river, has created several large islands. Some of these have gravel beaches that are the best places to find another of Teesdale's special plants. A gravelly beach on one of the islands looks accessible, more or less, by sliding down the riverbank and wading through shallow but fiercely flowing water. It's worth the effort – the edge of the beach is fringed by robust bushes of shrubby cinquefoil, covered in bright-yellow flowers.

Cultivars of this species are widely planted in gardens but wild plants grow in just two places in Britain. They can be found on a couple of tiny sites in the Cumbrian fells of the Lake District, although the best place to see them is a roughly ten-mile stretch of the Tees, centred on where I'm now standing. They also grow on the Burren, in Co. Clare in Ireland, another extensive area of Carboniferous limestone and one which, as we'll see on the last leg of our journey, shares further botanical treasures with Upper Teesdale.

Although shrubby cinquefoil is rare, it must be pretty tough to grow where it does – right at the water's edge along this stretch of the Tees – because here it must survive a phenomenon that we've already encountered further downstream: a

Tees roll. During a roll, the water level in the river here can rise as much as four or five feet in a matter of minutes – with no warning at all. Dramatic rolls like these are not as common as they once were, since Cow Green reservoir upstream on the Tees regulates the river level to a large degree. Even so, after heavy rain or snow melt high on Cross Fell, a lot of water flows over ground too sodden to absorb more, and spills into major tributaries of the Tees that remain unregulated. The vast flanks of Cross Fell channel so much water into the Tees that floods still pulse downstream to scour the riverbed and banks – and inundate the low, stony islands on which shrubby cinquefoil likes to grow.

Shrubby cinquefoil doesn't have particularly deep roots, but it is, well, shrubby, and this growth form may help anchor it against sudden floods. Where bent branches touch the ground, they take root and soon sprout a crown of new branches, some of which will, in turn, bend over and put down more roots. In this way, a big patch of shrubby cinquefoil has numerous anchor points, a built-in safety margin that allows it to survive on this unpredictable river.

The Teesdale Way continues north-east along the river, with the line of Holwick Scar, another outcrop of the Whin Sill, visible to the south, and beyond that the outlines of Holwick Fell. Alternating with pastures are lush meadows full of great burnet, their dark-crimson, oval flower heads floating in a sea of coarse grasses. This plant is more common in the North of England (although it can be found in the South), but it is characteristic of moist floodplain meadows, most of which have now vanished. So, the distinctive displays of great burnet are no longer a common site in many parts of the country.

In a few places, shallow valleys run down to the river and here, on the waterlogged valley bottoms, meadow gives way to bog. The sodden valley floors stand out in brighter shades of green or yellow, coloured by vibrant sphagnum mosses of various species and by sulphureous patches of bog asphodel. It's worth getting wet knees to have a closer look at the bog aspho-dels. Their yellow flowers surround brick-red anthers borne on furry yellow stalks that, from a more distant view, give the flower spikes a soft-focus appearance. Across the border to the west, they're sometimes called Lancashire asphodels – not a name that's likely to be adopted in rival Yorkshire. Here, their local name, 'moor-golds', is far more evocative. In the past, they've also been known as 'break-bones'. This odd name stems from a belief that a sheep grazing on bog asphodel will develop weak bones – one of those flaws in logic in which correlation is mistaken for causation.

Bog asphodels thrive in acid mires, where the peaty soil usu-ally has very low concentrations of calcium, a mineral that's vitally important for healthy bone growth. If sheep graze these areas, their diet is likely to be deficient in calcium, making their bones more prone to breaking. Bog asphodels are entirely blameless. That said, bog asphodels do contain some pretty nasty toxins and cattle sometimes exhibit symptoms of poison-ing if they're grazing boggy pastures.

The riverbanks are getting steeper now, rising to create sheer cliff faces as the Tees flows through a low gorge. The path meanders through gorse scrub and open woodland of scattered birch, ash and sycamore as it arcs away from the river, but a half-hidden sidetrack cuts back to the Tees, only to finish in a precipitous drop to the foamy water below. The path passes through dense scrub which deadens the sound of the river, but on emerging onto the open platform overlooking the river, the volume is suddenly cranked up. From here, there's a clear view to High Force, the biggest waterfall on the Tees, and even from this distance the roar conveys the sheer power of the water cas-cading over this high cliff of the Whin Sill. To get closer, I'll need to backtrack far enough to cross the river. On the north side, there's a path descending to the edge of the pool at the base of the falls, where it's possible to appreciate the true spectacle of High Force.

In the next chapter, on the final stage of our walk, we'll explore the natural history around High Force before following the river above the falls to its source, high on Cross Fell. Returning to the Teesdale Way, a low hill offers me a view of what lies ahead – ominous clouds drifting over the iconic outlines of the high fells – Mickle Fell, Cronkley Fell, Dufton Fell, Great Dun Fell and Cross Fell. After a long and tiring trek along the river, I'm once again reminded of Richard Watson's weekly commute. As I finish the last of the lukewarm coffee remaining in my flask, I rest on a conveniently placed boulder to survey a view similar to that which refreshed Watson near the end of his walk. He often paused at the top of Hardberry Hill, rising above Coldberry Gutter and the Coldberry Mine, to admire the beauties of Teesdale.

> Where sitting down, my wearied limbs to ease,
> I, looking back, survey the Vale of Tees
> What a majestic scene can be discern'd
> When to the far-off west the eye is turn'd;
> Grim mountains peaks in Alpine grandeur rise,
> Which in the distance seem to kiss the skies.
> First in the range, from hoary mist not clear;
> The outlines dim of high Cross Fell appear,
> With Dun Fell, Little Fell, and Meldon too,
> And nearer Mickle Fell's broad rocky brow.

The dim outlines that I, too, can see from where I sit are our eventual destination, but not before we get a closer look at High Force.

IO

Return to the Ice Age

*High Force to Cross Fell – Waterfalls, Weather
and Wildflowers*

Here's Cauldron Snout and High Force,
Magnificent sights worth surveying,
Wherever our footsteps may turn
New scenes are their beauties displaying,
And rare ferns and flowers adorn
Thy rocks, dells, and shady green bowers;
Thy sons are enduring and bold,
Thy daughters are fair as thy flowers.

RICHARD WATSCN,
'Lovely Sweet Vale of the Tees', 1834

The pitman-poet Richard Watson rejoiced in the scenic wonders of the Tees Valley in many of his poems, but it's clear from the lines above that he also understood at least something of the exceptional natural history of this place. Upper Teesdale is one of the top five botanical hotspots in the country, a showcase for individual species found nowhere else in Britain, as well as unique communities of ferns and flowering plants. We've already encountered shrubby cinquefoil on its exposed-gravel river islands, a plant known from just a couple of other places elsewhere in the British Isles, but this was merely an appetiser – the first in a profusion of special plants that we'll encounter on the final stage of our journey. These high fells and valleys, crags and wet grasslands are a 'must' on the bucket list of any serious botanist and have been so for several centuries. The singular nature of Upper Teesdale's flora was already well known to botanists during Watson's working life, in part because of the intrepid exploration and botanical expertise of another lead miner. More of his story later; first, we have a waterfall to explore.

At the end of the last chapter, I found myself on a sheer overlook on the south bank of the river, perched thirty feet above the roiling Tees, with a view upstream to the dramatic cascade of High Force. With a bit of precarious boulder-hopping, the crest of the falls is accessible from the Teesdale Way running along the southern bank, but to explore High Force more closely, and with at least a modicum of safety, we need to make a short backtrack to cross to the north bank over Soar Hill Bridge. Then, from the High Force Hotel, there's a path that descends through woodland to the base of the falls. The footpath runs through part of the Raby Estate, which charges a small fee for entrance, although many of my visits have been shortly after dawn in spring, far too early for the gatekeepers to be up and about – and at first light, I have the falls to myself.

Descending through woodland, the dense vegetation deadens the sound, but soon the roar grows louder, and occasional clouds of spray drift through the trees. To either side of the track there are spectacular clumps of water avens with hanging flowers of a subdued dusky red. For reasons I've yet to fathom, I've heard this plant called Indian chocolate,* but I'm not inclined to linger and contemplate this further. It's almost as if the power of the falls has energised the air itself and I quicken my steps down the sodden trail in anticipation.

The track emerges at the base of High Force, where, once clear of woodland, the sound is deafening. I've visited much bigger waterfalls around the planet, in Iceland (where a waterfall is also a 'foss') and in North and South America, but High Force is still special to me. As a young teenager, it was the first really big waterfall that I saw, and over the decades, the recollection of the awe I felt on my first encounter has remained just as vivid on all my subsequent visits.

It's not the highest waterfall in England. At just over sixty-five feet, it's beaten by both Hardraw Force in Wensleydale and Scale Force in the Lake District, both around thirty feet higher,† although High Force in full flow is much more powerful than either of these, and it stacks up against some of the most impressive waterfalls in Europe. The most spectacular one that I've seen is Dettifoss in Vatnajökull National Park in Iceland. On a winter's day, when all the rocks along the impressive gorge carved by the Jökulsá á Fjöllum River were painted in rime, the 300-foot-wide torrent plunging over 130 feet into the river canyon below presented a thrilling experience. At peak flow, nearly 20,000 cubic feet of water per second passes over these falls, making them the most powerful in Europe. However, with the Tees in full spate, over 10,000 cubic feet per second plunges

---

* I've heard of people in North America, where the plant is called the purple avens, boiling the roots with milk to make a brown liquid with at best only a vague resemblance to very bitter chocolate.
† Further south in Yorkshire, although hidden below ground, Fell Beck falls in an unbroken stream for more than 300 feet through the vast cavern of Gaping Gill.

over High Force, making it the largest waterfall in England by volume – and not too shabby a sight, especially as all this water plunges through the narrow channel of whinstone that confines the river at this point.

In full flow, High Force is actually excellent value for the very modest entrance fee, should you happen to have paid it. It's two waterfalls for the price of one. The river flows to either side of a tall central pillar of whinstone, creating two cascades, although on most occasions, visitors only see the southern channel flowing. Very rarely, the river level is high enough for the two falls to coalesce as one, as water flows over the central pillar as well. The last time visitors were treated to such a spectacle was in December 2015, during the torrential downpours brought by Storm Desmond.

In the past, High Force was subject to change without notice if a Tees roll swept down the river. A four- or five-foot rise in water level surging downriver from the high fells sounds spectacular, but it's so unexpected and so swift that a Tees roll is extremely dangerous. On 24 June 1880, two men, exhibiting more bravado than sense, clambered over the boulder-strewn river above the falls to reach the central rock pillar, from where they had a spectacular view. Unknown to them, a Tees roll was rolling towards them. Within a few minutes, the water level rose more than three feet, stranding them on the only remaining tiny island of rock in the middle of the swollen river. Seeing their predicament, other sightseers tried to rescue them by throwing ropes over the churning water. They succeeded in rescuing one of the men, but the other fell in and drowned. Had they known the Tees better, the two sightseers would have stayed where they were. In the wake of a roll, the water level drops almost as fast as it rises.

These men were not the first to be claimed by this stretch of the river – nor the last. In 1790, a young man crossing a ford at Leekwath, just below Middleton-in-Teesdale, was carried away by a sudden flood wave, and on several occasions the bridge at

Middleton-in-Teesdale was almost entirely submerged. Even today, in an era of over-zealous health and safety, people still die at High Force. Since 2010, three people have drowned here. A close approach is far safer from the bottom – and more rewarding. Great chunks of whinstone have broken off as the river erodes the softer rocks underlying the Whin Sill, undermining the harder rock until it can no longer support itself. The waterfall is in slow retreat upstream, while the remnants of previous incarnations lie scattered as moss-covered boulders around the edge of the deep pool at the base of the falls. It's not too hard to scramble over and around these boulders and to use the larger ones as cover to observe life around the plunge pool.

The first bird I see is a grey wagtail, almost hidden behind curtains of spray. It's flitting between boulders, wagging its tail, then dashing to the water's edge to grab an insect. It soon accumulates a beakful, then disappears under a rocky overhang at the river's edge – no doubt to a well-concealed nest and a hungry family. Downstream from the falls, a dipper is plunging beneath the roiling surface, where, despite the fast flow, it somehow manages to keep a grip on the gravelly bed with its clawed feet while it searches underwater for aquatic insects. It looks as though this bird, too, is feeding a family, and once it has gathered a hearty meal, it flies towards me. Dippers, with their stumpy wings, fly like bullets, and this one pays me no heed as it rockets towards the falls, then flies straight through the edge of the cascade. I don't think many predators will ever brave a raid on that nest.

Grey wagtails and dippers often nest very close together in the most secluded spots along a river. In a few instances, grey wagtails have even been seen to feed dipper chicks, although this isn't a case of inter-specific altruism. In the most closely observed example, the wagtails had to fly right over the dipper nest to reach their own nest a few feet further on. The shape of a bird passing over nestlings always elicits a loud begging response, which a parent bird – any parent bird – finds impossible to resist. Seeing a nest full of gaping maws, the grey wagtails responded by stuffing insects into the demanding dipper chicks. The wagtails' instinctive response had been fooled because the birds saw the right stimulus at almost the right place, but the dippers seemed to know exactly what was going on. As the wagtails brought more and more food to the dipper chicks, the adult dippers eased off on their own strenuous efforts, seemingly grateful for the unexpected help.

Watching the comings and goings of dippers and grey wagtails while cocooned in the comforting white noise of crashing water, I soon find myself in a meditative state and experience a powerful sense of timelessness. Nearly two hundred years ago, Watson doubtless witnessed this same scene, as perhaps did my own ancestors from Teesdale. Indeed, so too could countless generations before them. Through the centuries and along its whole length, the river has connected generations of farmers, miners, steelworkers and chemical workers alike. The Tees also ties all these seemingly separate past lives to our own modern existence and the world we've created. The river links all the disparate stories we've explored along the way into one narrative. The Tees flows through history as powerfully as it does through geography. It's only with great difficulty that I rouse myself from reverie, driven by a need to continue my journey and explore further upstream, above the falls.

Returning to the Teesdale Way, I follow the track past Keedholm Scar, back towards the southern side of High Force, through the tangled forest of scrub that this time is worth closer scrutiny. Many of the shrubs are ancient juniper bushes growing from artistically sculpted and gnarly stems. I'm entering the largest juniper forest in Britain, which runs (or ran) in patches from where I now stand at Keedholm Scar up to the summit of Cronkley Fell, several miles to the west, particularly where the Whin Sill runs close to the surface. Juniper was once so widespread here that 'jinifer' was collected for firewood, and juniper woodchips were used to fumigate local houses.

Junipers are conifers (one of only three native to Britain*) belonging to the cypress family. There are more than fifty species of juniper around the world, but only one species, the common juniper, *Juniperus communis*, grows in Britain. As a shrub or small tree, juniper is well suited to the tough conditions on the fells. In Tibet, juniper forests reach altitudes of over 16,000 feet, the highest treeline in the world. Common juniper is the most widespread species and grows right across the cooler parts of the Northern Hemisphere. Indeed, it's the world's most widespread woody plant – as well as the cornerstone of the gin-and-tonic-drinking classes. Gin isn't really gin unless it's flavoured with extracts from juniper berries.†

Neither cold temperatures nor battering winds worry juniper bushes much, but nibbling teeth are a different matter. Teesdale's juniper forests are dying, albeit very slowly. All the trees here are now ancient and there are no seedlings growing to replace them. There are just too many grazing mouths for any seedling to survive its early years. The culprits are not just domestic sheep – although, as we've seen, high densities of sheep have turned vast tracts of upland into biological deserts. In the absence of most of our original predators, rabbits are now so abundant on the fellsides that they, too, are having an impact, as, perhaps, are hordes of small rodents.

On top of that, since 2011 the plants here have been hammered by a pathogen (*Phytophthora austrocedri*) originating in related cypress trees in Argentina, but which has proved lethal to British junipers. Many of the bushes in the patch I'm exploring are already dead or dying and the same story is being repeated in many other places in Britain where juniper grows. To slow the spread of disease, some of Teesdale's juniper forests are now closed to public access and diseased trees are felled

---

* The other two being yew and Scots pine.
† Since juniper is a conifer, its aromatic 'berries' are actually female seed cones composed of fused fleshy scales.

and burned. In addition, various attempts have been made to restock the area with young plants grown from cuttings or from seed and caged to prevent grazing damage, but with limited success. It may be time for the gin-and-tonic-drinking classes to start worrying...

On this stretch of the walk, I've lugged my chest waders and a steel-framed dip net, since I want to see what aquatic insects live in the fast-flowing water here. For all of you that think Riverdance is a stage show of Irish music and dancing, you've clearly never watched an entomologist kick-sampling for river invertebrates. This involves wading into a couple of feet of water, then facing downstream and holding the net hard against the riverbed in front of you. Now, let the dance begin. You shuffle your feet backward and forward, loosening gravel and small stones, while trying very, very hard not to lose your balance in the pounding water – especially if you have an audience. Small invertebrates, and occasionally fish, disturbed from the bottom drift downstream and into the waiting net.

Emptying the contents into a white collecting tray, I can see that my energetic two-step has paid off. Several kinds of mayfly nymphs are swimming around, easy to spot against the white background. Mayflies thrive in fast-flowing, clear water, where they adopt one of two strategies to cope with the continuous pummeling. I can see several nymphs of a species called *Baetis roodani*, known to fly fishermen, somewhat more poetically, as the large dark olive. This species illustrates the first strategy. Its body is torpedo-shaped and beautifully streamlined.

Like all mayfly nymphs, it has a row of gills along each side of its abdomen, but the gills in this species are tiny, so as not to disrupt the elegant streamlining. In any case, oxygen concentrations are high in such cold, turbulent water, so it doesn't need large gills to extract the oxygen it needs. On several occasions, I've housed these nymphs in flow tanks that mimic their natural homes so that I can photograph them. Here, they happily walk around on the tops of stones, exposed to the flow of water, grazing on the film of algae.

There are several species of nymphs in my tray that illustrate the second strategy. The first that I recognise is *Ecdyonurus dispar*, a species that I've worked on extensively in the past. This is the autumn dun of fly fishermen, so called because, despite the name of this group of insects, it emerges as an adult in late summer. Unlike the large dark olive, the nymphs of this species are flattened, and their legs splay out to the sides to help them maintain a low profile. They live mostly beneath stones, although if they do crawl out, their flattened bodies allow them to remain in the boundary layer, a zone where friction with the stone's surface slows the water, thus allowing them to avoid the full effects of the current.

Many years ago, while I was still at the University of Bristol, I became fascinated with these nymphs. To film their method of feeding, I cultured a thin film of algae on a series of glass plates and persuaded these nymphs to feed on it, so that I could look through the glass with a binocular microscope to see how they used their complex mouthparts. Like all insects, they have three sets of mouthparts. From the two rear sets, the maxillae and the labium, sprout long, jointed appendages called palps, which these mayfly nymphs use to mow through growths of algae. Examining their mouthparts in a scanning electron microscope revealed that both sets of palps, which are flattened at the tips, are covered in an intricate pattern of ridges and stiff hairs, like rasps – perfect for scraping algae from stones.

Two other nymphs in my collecting tray resemble the autumn duns in being flattened and sprawling. A closer look will later reveal them to be *Rithrogena semicolorata* (the upright olive) and *Electrogena lateralis* (the dusky yellowstreak). However, the one species I hoped to find here seems to have eluded me. The upland summer mayfly, *Ameletus inopinatus*, has been recorded along this stretch of river and I've been keen to find it, since it's Britain's only Arctic-alpine species of mayfly. It lives in the northern half of the country, above a line between the Humber and the Mersey, but in the southern parts of its British range, it only lives high in the hills; further north, in Scotland, it descends to lower altitudes.

It can only survive in cold water, which is a cause for concern as global temperatures rise. Climate change has seen quite a few species of insects colonise Britain in the last few decades, but, as much as small red-eyed damselflies or tree bumblebees are welcomed as new arrivals,* so we will also lose species, such as summer upland mayflies or northern damselflies,† which are right on the southern edge of their range in Britain. (The same story, as we'll see shortly, is also true for birds.) Without adaptations for life in the fast flow, like those of the nymphs of large dark olives or autumn duns, upland summer mayfly nymphs prefer slower pools in upland rivers or calmer stream edges. However, despite performing my riverdance in all the likely places, I still can't find this species – and who knows how long I still have to track it down?

However, my efforts do yield the most impressive find of the afternoon – the huge nymph of a stonefly, *Dinocras cephalotes*. They reach over an inch in length and spend several years in the river before reaching full size and emerging as adults. These impressive creatures are predators, and studies in Ireland show them to be very fond of the nymphs of large dark olives, so this stretch of the Tees makes the perfect home.

My final prize is the larva of a caddis fly, *Rhyacophila dorsalis*, one of the green sedges of fly fishermen. As soft-bodied larvae, most caddis flies build protective cases consisting of carefully selected stones or precisely cut bits of vegetation, depending on the species. However, *R. dorsalis* is unusual – it's

---

* Small red-eyed damselflies were first recorded in Britain in 1999, tree bumblebees in 2001.
† This species, *Coenagrion hastulatum*, occurs in a very few places in Scotland, mostly around lochans on Speyside and Deeside.

a free-swimming species that doesn't bother with a cumbersome shelter. This endows it with much greater mobility than those species lugging heavy cases around, but at the cost of being a far more tempting target for the local brown trout.

In fact, most of the nymphs that I've found in my kick-samples feature in the diet of trout. When fully grown, mayflies and some stoneflies float to the surface, where they moult into their adult form. This process is remarkably quick, but even so, it leaves the nymph exposed to the keen eyes of fish. The autumn dun is an exception, very sensibly crawling onto a stone at the water's edge before moulting. Consequently, it's one of the river flies that's of least interest to trout – and therefore to fly fishermen. However, there are plenty of other river flies in this stretch of the river, so native brown trout thrive here. Unlike salmon and sea trout, they don't migrate downstream to fatten up at sea before returning upriver to breed, and that means they have the upper reaches of the Tees to themselves. There's no way past High Force for migrating salmon and sea trout, no matter how good they are at leaping waterfalls. The surging water is an impenetrable barrier, so the non-migratory brown trout in the higher stretches of the Tees don't face competition from these bigger, sea-fattened fish. They do, though, have to cope with fly fishermen.

To say that one of my closest friends and long-time film-making colleague Paul Reddish is an enthusiastic fly fisherman is a gross understatement; he's a man who thinks more like a fish than any fish that I've encountered – and that's the secret to being a good fly fisherman. Very often, trout focus almost exclusively on whichever insects happen to be emerging in the largest numbers, although not always. To keep fly fishermen guessing, trout occasionally ignore a mass emergence of juicy mayfly nymphs and feed instead on tiny blackfly. But in most cases, I'm told that the trick is to 'match the hatch'. You first need to tie a fly that mimics the real insect closely enough to fool the fish, and then you need to present it to the fish in such a way that its behaviour in the water also matches the real thing. Get all these things right and the chances are that a wily brown trout will still out-think you.

For one thing, it's not easy to persuade a fly, constructed from bits of feathers, to behave naturally, especially in fast-flowing water broken by lots of boulders. The line lying on the water gets pulled by the swirling currents and makes the fly move in unnatural ways – a sure way of tipping off a trout that something is not quite right. However, Paul tells me that this stretch of the boulder-strewn Tees is perfect for tenkara fishing, a form of fly fishing that evolved independently from that in the Western world.

It originates in the mountains of Japan, where the rocky streams look a lot like the upper Tees. Today, tenkara fishing employs ingenious telescopic rods that, folded down, fit into a coat pocket, yet extended can be ten feet or more in length. They taper to an incredibly narrow diameter and no reel is attached to the rod. Instead, an invisibly thin line snakes down from the tip of the rod. Those who've tried this technique (and it's rapidly growing in popularity in both Europe and America) say that the casting technique is easier to master than that of conventional fly fishing and such a long rod allows the fly to be placed precisely, perhaps in the lee of a big boulder, where brown trout love to rest up. Little or no line floats on the surface, so the fly behaves more like a genuine insect, hopefully enough to fool the shrewdest of fish.

Following the Teesdale Way upstream, the path crosses the river again at Forest-in-Teesdale before following Harwood Beck, a major tributary of the Tees. In the wake of the Norman invasion of Britain, William the Conqueror declared this part of the dale a Royal Forest, and it remained a hunting ground for many centuries for the likes of the Balliol family, although over time the size of the forest was reduced. One reason for the erosion of the hunting forest was slowly increasing settlement and farming in the upper dale. Not far from where I stand, Intake Farm is the perfect example.

Many of the farms here were originally 'intakes', parcels of land (often around thirty acres each) that had been 'taken in' from forest or moor. An intake is sometimes known as an 'unthank', another name which crops up in this area, both as a place name and a family name. Farms in the higher dales multiplied during the lead-mining era, since miners often needed to run a farm as well, just to make ends meet. Ideally these farms would be near the mines, which were often in more remote areas of the dales. Today, the landscape of scattered farms running up the valleys, giving way to wild fells above, is classically Teesdale – all the more so because every one of the farmhouses is painted bright white.

All this land is part of the Raby Estate, once owned by the Balliols and later by the Vane family, who, between 1833 and 1891, held the title of duke of Cleveland. A member of this family once became lost in bad weather in Upper Teesdale and only just managed to survive when he stumbled upon a farmhouse. Such dangerously bad weather is all too frequent in this part of the world, so, to forestall an even worse outcome in the future, the then-duke of Cleveland ordered all the farmhouses on his land painted white, to make them stand out as beacons to lost travellers.

The Teesdale Way now leads me west, across Harwood Beck, to rejoin the Tees a little way to the south-west of Langdon Beck, home to a fine pub and a refreshing end to a day's tramping over the fells. From here, the river runs through a steep, craggy valley which rises to Widdybank Fell in the north and to Cronkley Fell in the south. These fells are part of the Moor House-Upper Teesdale National Nature Reserve, one of the first National Nature Reserves to be designated, in 1952. Originally, Moor House, centred on Cronkley Fell, and Upper Teesdale, centred on Great Dun Fell to the west, were separate reserves, but they were joined at the end of the last century to create England's largest – and highest – terrestrial National Nature Reserve, covering around thirty square miles of Pennine grassland and moorland.

For any botanist, the names 'Widdybank' and 'Cronkley' should be enough to quicken the pulse. In spring, the intense-blue stars of spring gentians dot the grassland. In wetter patches, tiny pink heads of bird's-eye primrose outshine any blousy garden variety. Drifts of mountain pansies, with their outsize petals in mixtures of blue and yellow, look like swarms of tropical butterflies fluttering over the fell. The rather more conventional-looking Teesdale violet blooms in a few spots on Widdybank Fell, one of only four places in which this plant grows in Britain.* Teesdale sandwort is even more special. The half-dozen or so gravelly flushes on Widdybank Fell in which it grows are its only sites in the whole of Britain. And this is just a small fraction of the Teesdale Assemblage, the unique flora of Upper Teesdale.

As remote as the North Pennines are, the flora here has been attracting the attention of naturalists for many centuries, including some who were celebrities of their day. John Ray (1627–1705) was one of the earliest in that great English tradition of parson-naturalists, living at a time when a career in the Church left ample time to pursue natural

---

*   This plant also grows on Long Fell and Arnside Knott in Cumbria, and at Ingleborough in Yorkshire. Elsewhere, it grows from Northern Europe all the way across to the Russian Far East.

history as well. For Ray, though, as for many that followed in this tradition, the pursuit of natural history was as much a quest for an understanding of God through his works of Creation as it was an exploration of science. Indeed, Ray promoted these ideas in his book *The Wisdom of God Manifested in the Works of the Creation*, published in 1691.

Nearly a decade and a half before this publication, in 1677, John Ray visited Upper Teesdale to meet with Ralph Johnson (1629–95), who became a great friend and who, like Ray, split his time between the Church and nature. Johnson's day job was as vicar of the parish church at Brignall, a small village near the River Greta a few miles upstream from Greta Bridge. However, Johnson seems to have spent a lot of time exploring the upper reaches of the Tees, where he was able to show Ray the specimens of shrubby cinquefoil that he'd discovered – at the time, a plant still new to science, since Johnson hadn't published his finds.

Johnson also discovered dwarf birch and Scottish asphodel on Widdybank Fell. It's likely he also knew of spring sandwort, northern bedstraw and alpine bartsia, but, as with his other finds, he didn't publish these discoveries. The lack of published material makes the early history of the discovery of Teesdale's flora something of a mystery. Nonetheless, the historical botanist Frank Horsman has explored sources as varied as legal letters and surviving herbaria in forensic detail in an attempt to shed further light on the identity of these botanical pioneers. He suggests that Johnson's work may have been followed by that of an even more shadowy figure, Christopher Hunter (1675–1757). Hunter was an antiquarian but was also an acquaintance of Johnson, who likewise botanised in Upper Teesdale. He is known to have transcribed Johnson's papers, including a list of fifteen plants that Johnson had found in Upper Teesdale. However, beyond this, he's virtually unknown as a botanist and we're unlikely now to gain many more insights into his role in discovering Teesdale's rare plants.

After Hunter, botanical exploration fell into a lull for several decades, until a surgeon from the Scottish Borders moved
into the area in 1783. William Oliver (1761–1816) travelled
widely through Upper Teesdale on call and doubtless picked
up snippets of botanical interest from local farmers and miners,
who would certainly have been familiar with the more obvious
flowers, such as spring gentians. Sowerby and Smith's *English
Botany* of 1798 tells us that local people knew this plant well
'by the name of Spring Violet, as it copiously enamels that
country at a time when no other flower enlivens the dreary
scene'.* There were even competitions to see who could produce the best display of gentians, by covering a ball of clay or
moss with flowers and showing off the creation on a window-
sill for all to admire.

Oliver's discoveries may also have passed unpublished and
unnoticed were it not for his meeting with Reverend John
Harriman (1760–1832). Harriman began life as a medical
student but switched to the Church, and, in 1787, became
curate at Bassenthwaite in the Lake District before moving to
Barnard Castle in 1796, where he met Oliver. Harriman was
another cleric with time on his hands for plant-hunting but,
more importantly, he was a Fellow of the Linnean Society, and
therefore well connected with the mainstream botany world
of the day.

Harriman knew Edward Robson, a botanist living in
Darlington, and sent him Oliver's specimens. In turn, Robson
circulated the findings to others in the Linnean Society, including Smith and Sowerby, who were preparing their botanical
magnum opus. Harriman is often considered to have made a
significant contribution to the story of the Teesdale flora, but
Horsman's evidence suggests that most of the discoveries were
Oliver's. Oliver also worked with another intriguing character

---

* This massive work was a 36-volume set, issued in 267 monthly parts over 23
years from 1790 to 1814. Not really a handy field guide.

in this story, who may also have been over-credited – John Binks (1766–1817). Unlike the assorted clerics and medics who had both time and money to travel in Upper Teesdale, Binks was a lead miner, although he only worked a four-day week in the mines, which allowed him time for botanising. He became familiar with the botany of Upper Teesdale through collecting plants with medicinal properties (simples) for local herbalists as a way of supplementing his income as a miner.

Binks undoubtedly discovered some of the rarities, such as yellow saxifrage in Baldersdale, where he described the bogs as yellow with this plant. For other plants, it's often hard to know whether it was Binks or Oliver who made the original discovery. Again, Frank Horsman feels that most of the credit should go to Oliver. Perhaps this perceived injustice stems from Binks's later acquaintance with the influential Backhouse family, who we last met in Darlington as backers of the Stockton and Darlington Railway. Father and son James (1794–1869) and James (1825–90) Backhouse both explored Upper Teesdale. James Senior was a nephew of Edward Robson and would therefore have been keenly aware of the remarkable discoveries being made along the upper Tees. With John Binks as a guide, the Backhouses explored Teesdale and collected many of the rarities. In fact, the Backhouses became such frequent visitors to Teesdale that one of the rooms at the High Force Hotel became known as 'Mr Backhouse's room'. In their notes of these trips, they often credited Binks with the original discoveries, even though many were first found by other botanists. In any case, I still admire John Binks as a working-class man who, despite a physically demanding and poorly paid job, still had the motivation to make long treks into the hills in search of plants.

Although the Backhouses made some original discoveries themselves, by the time they were exploring the dales in the early nineteenth century, many of the rarities had already been described. In 1798, Edward Robson printed the catchily titled *Plantae rariores agro Dunelmensi indigenae* (*Rare Native Plants of County Durham*), which featured spring gentian, alpine bartsia, Scottish asphodel and mountain avens. A few years later, Nathanial Winch published *The Botanist's Guide through the Counties of Northumberland and Durham* (1805), the first flora to feature Teesdale in some detail. The second edition of Richard Garland's *A Tour of Teesdale* publicised new discoveries, such as spring gentians, mountain avens, Teesdale violets and bog orchids.[*] In the same year, Teesdale tourism received a massive boost with the publication of Walter Scott's epic poem *Rokeby*, and visitors flocked to this remote valley, eager to experience the wild, romantic landscapes described by Scott along with its unique flora. In these early decades of the nineteenth century, Napoleon Bonaparte also contributed to the popularity of places like Teesdale as tourist destinations. The Napoleonic Wars (1803–15) put a temporary halt to the traditional Grand Tour of Europe for the wealthy, who turned instead to the more remote areas of Britain. As visitor numbers to Upper Teesdale soared, botanical tourism grew into something of a local industry.

John Binks became a botanical guide for visiting naturalists, while the landlords of both the High Force Hotel and the Langdon Beck Inn also got in on the act. Soon, Teesdale was on every botanist's itinerary – even the most famous. Joseph Dalton Hooker (1817–1911), one of the most important botanists of the nineteenth century, became director of Kew Gardens in 1865. In the summer of that year, however, he was staying at the High Force Hotel while his wife recovered from a recent miscarriage, and while he studied the glacial morphology of the landscapes of Teesdale – although he seems to have found the local cuisine considerably more attractive than the high fells. He wrote to his friend Charles Darwin:

---

[*] Richard Garland. *A Tour of Teesdale Including Rokeby and Its Environs.* 2nd ed. Thomas Wilson and Sons. 1813.

Dear old Darwin

This is a wonderfully pretty place, especially the banks of the river. [T]he moors are loathsome & the hills low broad rock-less treeless & contemptible. I am studying the moraines all day long with as much enthusiasm as I am capable of after laying in bed till 9 eating heavy breakfasts & looking forward to dinner as the summum bonum of existence.[*]

Joseph's father, William Hooker, director of Kew Gardens before his son, had also visited Upper Teesdale in 1817, and Hookers senior and junior corresponded with Backhouses senior and junior, so the special Teesdale flora was well known to both William and Joseph Hooker. Even so, and despite the area being so well trodden by generations of keen-eyed botanists, new plants continued to be discovered.

In the 1950s, little less than a century after Joseph Hooker explored Teesdale, Margaret Bradshaw arrived in the area and began what eventually turned into a seventy-year botanical study. It's still going on. At the time of writing, Margaret is ninety-eight years old, and still gets around Teesdale on horseback. Almost as soon as she arrived, Bradshaw discovered large-toothed lady's mantle in some of Teesdale's hay meadows, a plant never seen before in Britain. Today, its only known sites – apart from a single record in Weardale, to the north – remain in Teesdale. There are many kinds of lady's mantles, which are notoriously difficult to distinguish from one another, but Bradshaw is an authority on these plants and knows what to look for. Such a discovery makes you wonder what else is lurking in the meadows and high fells of this remote area, just waiting for someone with sufficient knowledge to reveal it.

As if to prove this point, in 2019, the team set up by

---

[*] Letter from Joseph Dalton Hooker to Charles Darwin, 13 July 1865, High Force Inn, Middleton-in-Teesdale.

Bradshaw to study the Teesdale flora discovered Ostenfeld's eyebright growing on Little Fell and Mickle Fell. Eyebrights are another fiendishly difficult group of plants to identify, since their taxonomy is complicated and frequently changes as new classification techniques evolve. This discovery seems to be the first record not just for Teesdale, but for the whole of the North Pennines. Elsewhere, this eyebright is scattered across northern Scotland and on mountains such as Helvellyn in the Lake District.

Most botanists regard Upper Teesdale as one of the top five botanical sites in the British Isles. Its rivals are the Burren in Co. Clare, Ben Lawers (Beinn Labhair) in Scotland's Southern Highlands, the Lizard in Cornwall and the Brecks of East Anglia. These places couldn't be more different; all have their own unique histories and ecologies, and different assemblages of plants, although Teesdale does share some botanical affinities with Ben Lawers and the Burren.

However, it's not just the existence of its rare plants that puts Teesdale on the botanical map. The *communities* of plants here are unique, even on a global scale. Plants with many different distribution patterns nestle side-by-side along the upper Tees. There are Arctic-alpines, such as alpine bartsia, which beyond Teesdale, as the name suggests, occur in both the Arctic and the high mountains of Central Europe. There are purely Arctic species, such as the somewhat confusingly named alpine foxtail, a kind of grass. The famous spring gentians are widespread in the mountains of Central Europe, while bird's-eye primroses are plants of northern mountains right across Europe and Asia. Then there's horseshoe vetch. Among all these hardy mountain and Arctic plants, horseshoe vetch is a southern species, commonly found on much warmer limestone grasslands in southern Britain and further south through Europe. Nowhere else in the world does such an eclectic mix of plants grow together.

Why is Teesdale botanically so special? The short answer is that we don't really know. The long answer, however, is a little more enlightening. To understand at least part of the reason for the existence of Teesdale's special flora, we must travel back to the end of the Ice Age. When the last glacial phase of the Ice Age began to loosen its grip, around 15,000 years ago, the ice sheets still extended as far as the southern half of Britain and the land beyond the ice was frozen tundra, although it was not without botanical interest. Travel to the tundra of the high Arctic today and you'll get a sense of what the land from Kent to Cornwall once looked like. Even better, you can catch a train to the Ice Age in southern Norway. The line from Oslo to Bergen crosses the Hardangervidda, the high plateau of Hardanger, which lies above 3,000 feet. It's the largest such plateau in Europe and high enough for the land to be a little piece of the Arctic nearly 500 miles south of the Arctic Circle.

Jump off the warm train at the little station of Finse and you'll find tundra plants before you even leave the platform. Growing in grassy patches by the station buffet are deep-blue alpine gentians, a plant that you'll need to trek up to the summit of Ben Lawers to see in Great Britain. It's not a long walk from the station to the snout of Hardangerjøkulen, one of the largest glaciers in Norway. The journey crosses rugged, stony ground with scattered sedges and grasses and a host of colourful tundra plants. Several of these plants are the same species that I've seen in Upper Teesdale, although many more no longer occur in Britain, even though they were common there at the end of the Ice Age. So, the Hardanger Plateau today provides a glimpse into Teesdale's distant past (apart from the railway line, of course).

The luxuriant lichens of the Hardanger Plateau are grazed by herds of wild reindeer. During the Ice Age in Britain, reindeer like these foraged over the South Downs, where they were joined by musk oxen, huge, hairy beasts that fed on grasses, sedges and herbs and helped keep the landscape treeless as the climate slowly warmed. They still lumber around the high Arctic today, but sadly, no musk oxen graze the Hardangervidda. However, in central Norway, they've been reintroduced to Dovrefjell, a range of mountains within hiking distance of Fokstua station. Natural history trips through Scandinavia are perfectly possible – and more eco-friendly – by train, although Fokstua station feels like it's in the middle of nowhere. Standing there, watching the train disappear into the vast landscape with nothing but a few days' supply of food and a small tent, was slightly nerve-wracking.

I travelled there to see yet more Arctic plants, but most of all to track down the reintroduced musk oxen. Over the next few days, I was lost in wonder as I discovered tiny white Arctic orchids and colourful saxifrages – but not a sign of a single musk ox. This is a huge area but, even so, I felt that it was hard to mislay something the size of a musk ox. Eventually, I headed into a small village in the valley to enquire further – and it turned out that Dovrefjell's musk oxen herd had met with disaster.

In response to a threat, such as a marauding wolf pack, musk oxen gather in a tight phalanx, the big males at the front, protecting females and calves behind. They present a defiant barrier of bony heads carrying large, sweeping horns – one that only a desperate predator would want to brave. Two days before my arrival, a huge thunderstorm had swept over the mountains and to weather the storm on an exposed mountainside, the musk oxen had huddled up in tight formation. Unfortunately, the herd was struck by lightning, and many were killed. However, the enterprising people of Fokstua had hauled the bodies off the mountain and turned them into burgers in a local café. Still to this day, my only sighting of a musk ox has been between two halves of a bun. It was, however, delicious.

As the climate warmed between 15,000 and 12,000 years ago, the big grazers of Britain's Ice Age tundra had a slower and somewhat less dramatic decline than Dovrefjell's musk oxen. In Britain, these animals, and the Arctic flora on which they grazed, moved north with the retreating ice and we have a good

picture of these changing times thanks to the science of palynology – the study of ancient pollen grains.* Pollen is covered in an
almost indestructible coating which often bears characteristic
spines or surface sculpting. Plants produce vast clouds of the
stuff, most of which ends up on the ground, where, owing to
their tough coats, the grains may survive intact for thousands of
years. Peat deposits, which provide ideal conditions for pollen
preservation, record the changes in abundance of different
kinds of plants. The distinctive structure of pollen grains allows
many of them to be identified to individual species, although
some, such as pollen from different willows, is so similar that
palynologists can't distinguish between the many kinds of willows that thrived from the Ice Age to modern times in Britain.

Such analyses of pollen from this period reveal that, at the
end of the Ice Age, Teesdale's rarities, many of which today are
plants of the far north or of high mountains, grew widely over
most of tundra Britain. Later, as the climate continued to warm,
these tundra plants were replaced by plants that thrived in the
more temperate conditions. However, the pollen record shows
that these plants never left Upper Teesdale. The fells and crags
provided a sanctuary for these Ice Age refugees.

How did the Teesdale Assemblage survive here – and, in the
cases of a few plants, nowhere else? What did Teesdale offer
these plants? First, there was the Whin Sill. At the end of the
Carboniferous Period, as molten magma welled up from deep
below and flowed through the overlying rocks, the heat chemi-
cally altered the rocks. Where the magma cooked the limestones laid down earlier in the Carboniferous Period, it created
a form of marble – but unlike more familiar kinds of marble,
these baked rocks weather easily and crumble into a fine gravel
with the texture of granulated sugar – hence they are known
as sugar limestone.

---

* Palynology also includes the study of other tiny particles of both organic and
inorganic origin – it literally means 'the study of dust'.

Areas underlain by porous sugar limestone often remain open, with many bare patches of limestone gravel. All the Teesdale rarities are plants that only thrive in open conditions, so the sugar limestone creates a perfect home for them. Today, a great many of Teesdale's special plants are found on the sugar limestone, which outcrops as narrow bands along the flanks of Cronkley and Widdybank Fells. Teesdale sandwort is so particular that every one of its few sites are on bare sugar limestone gravel. But the pollen record tells us that as the climate warmed further, trees began to march northwards. Soon, most of Britain was covered in forest, including the bleak slopes of Cross Fell, the highest point in the Pennines. Even the summit had a sparse covering of scattered, stunted trees, although it probably remained open enough for many of Teesdale's rarities to survive. But how did plants that thrive in open conditions survive in the rest of the wildwood?

We've already seen that the wildwood was not the vast, impenetrable forest of European fairy tales, but a mosaic of forest, scrub and open grassland, a dynamic ecosystem maintained by the big grazers and browsers, such as the aurochs and horses that eventually replaced the reindeer and musk oxen. Large patches of Teesdale probably remained open, especially on the thin, dry soils of the sugar limestone. In addition, the erosion-resistant Whin Sill also created the cliffs and rock ledges that characterise the upper valley of the Tees. Such places were never overshadowed by trees. The Tees itself, with its frequent powerful floods, scours the banks and sweeps away any trees colonising the gravel islands. So, many of the characteristics of the Tees and its valley that we've encountered in our journey combine here to create a place where a unique flora has survived since the end of the last Ice Age.

However, the climate of Upper Teesdale has also played a pivotal role both in the survival of the Ice Age refugees and in the creation of its unique plant communities. The area is unusually cold, which helps keep trees at bay and maintain the open landscapes that the Teesdale rarities need. As I stand on the windswept flanks of Widdybank Fell, looking towards the distinctive skyline of Cross Fell and Great Dun Fell, I can see an epic skyscape evolving in front of my eyes. Purple clouds shift against each other and grey curtains of rain part to reveal shafts of sunlight that rake over the dark fells like searchlight beams. It's the middle of May but it certainly doesn't feel like that. I'm reminded of a comment from the nineteenth-century *Gentleman's Magazine* which described Cross Fell as being buried in snow for ten months of the year and in cloud for eleven months. That's a little exaggerated, but snow cover has often lasted for three months, and frosts can occur in every month – so this place really is a little bit of the Ice Age in modern Britain.

When the Moor House National Nature Reserve was set up, the warden's house was the highest occupied house in England and was often cut off in winter. One winter, the warden was isolated here from 21 December 1979 all the way through to 21 April 1980. It certainly gets cold here, and often very quickly, thanks to another local phenomenon – the Helm Wind.* This fierce, chill blast is the coldest wind in Britain and originates in the Vale of Eden on the other side of the fells from where I stand. The howling wind has been known to blow down buildings standing in its path.

It's a 'katabatic' wind that originates as a mass of air that's been pushed to the top of the mountains. Here, it cools and becomes denser, and therefore begins to descend from the peaks (katabatic derives from the Greek for 'descending'). Katabatic winds can be ferocious, reaching hurricane speeds, but thankfully, though destructive, the Helm Wind has never achieved such fearsome power. As the air mass descends, it warms up,

---

* The Helm Wind gets its name from the fact that it is often accompanied by a cap (a helmet, or helm) of cloud flowing off the high tops. The most famous such helm is formed by the clouds that cap Table Mountain above Cape Town in South Africa.

although when the Helm Wind rips through Upper Teesdale it's still bitterly cold. A more famous katabatic wind, the Santa Ana in southern California, heats up so much as it descends that it becomes unbearably hot – it's even been known to drive people mad.

With such unpredictable weather on the high fells and the destructive, cold blast of the Helm Wind, it's hardly surprising that earlier inhabitants of Upper Teesdale called Cross Fell 'The Fiend's Fell', an abode of evil spirits. Six hundred years ago, monks from Hexham Abbey erected a cross on the flanks of the fell to exorcise the demons. To judge by the weather brewing in front of me today, they didn't succeed. However, the gathering dark clouds are a dramatic demonstration of the Arctic conditions up here that have allowed the Teesdale rarities to survive. In the face of the intensifying storm, I make a strategic retreat to the pub at Langdon Beck, ready to push on along the river when the weather improves.

The following day, the sky is as blue as the spring gentians studding the green turf on the fells. Walking along the valley of the river, edged with the cliffs and tumbled rocks of Falcon Clints, I soon see the distinctive white flash of a wheatear. This buff, grey and black bird, with a highwayman's black mask, arrives here in the spring to make its home among scree and boulders. It's an easy bird to identify, with its white rump, framed by a black 'T' – indeed, its name comes from the Old English for 'white arse'.

Somewhere among the boulders, someone seems to be busy typing on an old-fashioned typewriter – the calls of a ring ouzel. Ring ouzels are rare upland relatives of the familiar blackbird, distinguished by a broad necklace of white feathers. They also differ from blackbirds in that all our ring ouzels are migratory, heading south to North Africa and southern Spain for the winter.

For me, though, the real voice of the high fells belongs to the golden plover. We met these birds earlier, in their winter flocks in the lowlands, around the Tees Estuary. At that time of year, they're dressed in more subdued shades, but now they're decked out in their breeding plumage, with jet-black bellies and face masks and golden spangling on their backs. Their calls, echoing over the fells, are heart-rending and plaintive low whistles. They breed on most upland areas in Britain but are most abundant on these North Pennine fells. In the past, they were joined here by an even more beautiful wader – the dotterel.

Dotterels are true birds of the tundra, nesting only on the exposed tops of mountains in the south of their range, although nearer to sea level in the far north. In the middle of the nineteenth century, some fifty to seventy-five pairs nested on the uplands of the North Pennines and North Wales, where breeding was recorded annually up until 1927. After this, numbers declined and soon dotterels disappeared as breeding birds from England and Wales.* Now, their main breeding population is on the high tops of the Cairngorms in Scotland, the last patch of true tundra remaining in these islands after the Ice Age ended.

Dotterels still call in on Teesdale on their migration north in spring, and occasionally on their southbound journeys in autumn. They travel in parties called 'trips' that pause at traditional rest stops on their long journeys. These habitual stopovers mean that, with a bit of persistence and patience, you can all but guarantee seeing dotterels in one of these places in spring. I've never seen them in Teesdale, but there's a regular stopover at Danby Beacon, an exposed, high area of the North York Moors, to the south of Middlesbrough.

Here, I've seen flocks of a dozen or so dotterels pottering unobtrusively among the heather. They're colourful birds, each with a black cap fringed with a bright-white eyestripe and a narrow white breast band that separates a grey chest from a chestnut belly. The bottom of the belly grades to such a deep

---

* Intensive fieldwork for the *Bird Atlas 2007–11* recorded one breeding pair in northern England.

colour that they look like they've been dipped in dye. I can only describe these birds as endearing. They will rarely do more than acknowledge you with a quick glance, even when you get within a few feet of them. I can sit in the heather, surrounded by their quiet pipings, almost as part of their flock. I don't know where the collective noun 'trip' for a dotterel flock comes from, but in my mind it's because you can almost trip over them. Sadly, their charming and confiding nature has played a part in their decline.

In the nineteenth century, there are reports of trips of thousands of birds, which drew equally large flocks of hunters – hardly sport with such a trusting creature. They simply saw the birds as stupid, since they could so easily be enticed into nets. Earlier in our journey, we saw that the wholesale slaughter of great crested grebes, merely for their feathers, almost wiped out these birds in Britain. Dotterel feathers were also highly prized; those deeply coloured belly feathers were eagerly sought-after by fly fisherman. The North York Moors drew so many 'hunting' parties that local inns prospered during the hunting season. The Dotterel Inn at Reighton, near Filey on the North Yorkshire coast, is a reminder of those days.

However, hunting isn't the only reason for the dotterel's retreat north. Like many species for which Britain lies at the southern edge of their range, our warming climate is forcing dotterels to move both further north and higher into the mountains in search of the cool climate and open mossy vegetation that they need. One study in Scotland suggests that dotterels are having to move upslope at about seventy-five feet per decade. Clearly, there will come a point where they can retreat no further, and we'll lose these delightful birds from our islands.

Other mountain birds face similar problems. Sitting patiently among the boulders of Falcon Clints, I finally spot a ring ouzel with a beakful of earthworms. It disappears quickly into the jumble of rocks to its well-hidden nest. I get a real thrill from seeing this bird, if only briefly, since it's getting much harder to find. Their population in Britain has been declining since the first half of the twentieth century, but more recently the decline seems to have accelerated. From the early 1990s to the end of the decade, ring ouzels suffered a 58-per-cent drop in their numbers. There are now only around 7,000 breeding pairs in the whole of Britain.

The elusive ring ouzel is a tricky bird on which to gather hard data, but the latest studies suggest that its decline is being driven by climate change, both here and in its winter retreats. Birds from Britain winter mostly in the Atlas Mountains of Morocco, where they feed on the berries of the local species of juniper. The changing climate has reduced spring rainfall in this region, which in turn reduces the late-summer crop of berries. Once back in the high fells, ring ouzels rear their chicks largely on earthworms, just as the one I've glimpsed has been doing. Warmer and drier springs and summers may limit the availability of food for their families, although climate change is probably working against these birds in lots of different ways.

Leaving the plovers, wheatears and ouzels to carry out their parental duties in peace, I reach the confluence of Maize Beck with the Tees. Then, rounding a jumble of boulders, I find myself at the bottom of Cauldron Snout, a cascade of white-water tumbling through a chasm cut into the Whin Sill. More a cataract than a waterfall, at 600 feet long, Cauldron Snout is reckoned to be the longest in England. It's also reputed to be haunted.

Local legend has it that the sorrowful laments of the Singing Lady can be heard above the roar of water and the shade herself is often glimpsed seated on a boulder. The Singing Lady is the ghost of young Victorian girl who drowned herself in the cataract after being parted from her lover, a lead miner who was camped far away from his home at a nearby 'shop', as he worked the mines around Widdybank Fell. Climbing up the side of Cauldron Snout, I see and hear no trace of the lady, although my concentration is almost entirely focused on the damp crevices and fissures where lush ferns and mosses thrive.

At the top of my climb, my reverie is shattered when I'm confronted with the vast concrete wall of the dam that holds back the waters of the Tees to create Cow Green Reservoir. It's a stark contrast to the untamed beauty I've been tramping through – defiling the wild landscapes of the dales. I remember well the controversies surrounding the building of this reservoir, which was filled in 1970, flooding a significant part of the hallowed botanical turf of Teesdale.

Prior to 1967, when the Tees Valley and Cleveland Water Board chairman pressed the plunger of a detonator to blast a crater out of the ground above Cauldron Snout – an act that appropriately symbolised the wanton vandalism of such an important natural history site – the Tees above Cauldron Snout was a unique stretch of the river. Unlike the fast-flowing, boulder-strewn reaches along which I've been walking, this stretch of the river flowed more slowly through a swampy basin, and in places formed a long pool known as the Weel. Apart from possessing a rich and intriguing natural history, this reach of the upper Tees was home to another mythological figure, far older than the Singing Lady. The Weel was the haunt of Peg Powler.

Peg was a 'grindylow'. For those who've read the Harry Potter novels, the term will be familiar – green-complexioned water spirits with long fingers and sharp teeth that lure people into deep water to drown them. However, J. K. Rowling didn't just conjure these from her fertile imagination. Grindylows* are mythological water spirits from Yorkshire and Lancashire. Peg Powler was similarly said to have green hair and the rather inconsiderate habit of luring people into the river to drown them. Her daughter, Nan Powler, lives in the River Skerne, the tributary of the Tees that runs through Darlington. These

---

*   Grindylow is probably cognate with Grendel, a character in *Beowulf*, the Old English epic poem.

water spirits living along the Tees take us back to a time when rivers were such important features of life, providing transport, food and water supplies, that they were seen as sacred. At their sources, emanating from the ground as mystical springs, rivers were seen as gifts of goddesses, to be treated with more respect and reverence than has been the case along the Tees. Unfortunately, Peg Powler was unable to save her home.

In the 1950s and 1960s, industries were expanding on Teesside and needed ever-larger supplies of water. The Water Board's chairman said at the time: 'we have not built it [Cow Green] for any other reason than to assist industry with its increasing needs for production'. However, it turned out that Cow Green was the solution to a problem that never was. By the time it was filled, Teesside's industries had slipped into decline and the demand for water slowly disappeared. At the time, the reservoir was strongly opposed by conservationists, including Margaret Bradshaw, who by then knew as much about Teesdale's flora as anyone, and by David Bellamy, a charismatic botanist who, through his many television appearances, made botany sexy in the 1970s. Protests had begun in 1957, but these efforts were in vain – short-term gain, as always, trumps longer-term thinking.

The only small consolation was that ICI, one of the industries that would benefit from the water supply, gave £100,000 to create a fund for botanists to study the area that was to be inundated – so at least we would know what we had just destroyed! Margaret Bradshaw had only a short time to assemble a team and carry out the survey work, much as Peter Evans had to do when land reclamation projects began in the estuary. The Tees was being assaulted from its headwaters to its lowest reaches.

In the end, a unique stretch of the river, along with a tenth of the population of Teesdale violets and nearly half of the habitat of early spring sedge, another rare plant, were all submerged beneath a pointless reservoir.* And Cow Green continues to affect life in the Tees, long after it was filled. A large, deep lake where none existed before has provided suitable conditions for huge numbers of planktonic crustaceans which are regularly flushed into the river, where they change the river's ecology. The dam regulates the downstream flow of the river, evening out variations and making big Tees rolls a thing of the past – almost.† Gravel and sand settle out in the calm waters of the reservoir, reducing supplies downstream and further shrinking the spawning beds of salmon and trout. In addition, the more

---

* Shortly after Cow Green, Kielder Water on the headwaters of the North Tyne in Northumberland was filled, and a complex network of tunnels was constructed to link the Tyne, Wear and Tees. Kielder Water is the largest reservoir in the UK by capacity, and was also built to supply industries in the North-East.
† Maize Beck still gathers runoff from a vast area of the fells, which can quickly raise water levels in the Tees.

uniform flow in the river prevents scouring during periods of high water, and this in turn creates a more stable riverbed and a thicker growth of algae. Such growths wipe out the flattened mayfly nymphs that I found further downstream. Where once the Tees resembled its unregulated tributary Maize Beck, now the two rivers have very different ecologies.

As is the case with the Tees Barrage, fifty miles downstream, Cow Green has far-reaching effects along the river. Curly pond-weed, which forms thick stands in slower rivers, has moved fifteen miles upstream after impoundment at Cow Green. This species, along with whorled water milfoil, is now found upstream of the confluence with the Skerne at Darlington, so the dam's effects reach even further than those of the barrage. Most of the river has been radically altered by these two developments alone.

There's no public footpath upstream along the Tees from Cow Green, but an old miner's track along the flanks of Herdship Fell allows me to walk a mile or two upstream. Dropping down first to the heavily eroded shoreline of the reservoir, I flush a ringed plover. These birds nest on open patches of gravel, so the barren shoreline of the lake suits them well. In fact, this species only moved into Teesdale as a breeding bird after the completion of the reservoir. They're elegant birds, marked with a black collar and mask set against a clean white belly and neck ring, and they are always a pleasure to watch, so I grudgingly admit that this is one bonus of the reservoir. Their nests are very hard to see and very easy to trample. I don't think I'm really close to a nest, since, like many waders, ringed plovers have a very convincing 'broken wing' act designed to lure predators away from eggs or chicks. Even so, it's best not to risk it, so I set off along the miner's track for a short detour.

The footpath, now marked on my Ordnance Survey map as the Pennine Way, crosses the river just below the dam, then heads off across country, parallel to Maize Beck, past Moss

Shop and nearby Maizebeck Shop at the heads of old lead mines. Was one of these the working-week home of the Singing Lady's lover? Eventually, the path drops down to the bank of Maize Beck and then up to the edge of a dramatic valley, High Cup Gill. A notch at the head of this valley, High Cup Nick, offers a spectacular view, though it's hard to even stand up in the howling wind, let alone appreciate the scenery through watering eyes. However, this track is taking me away from the final part of my journey along the Tees. The source of the Tees lies high on the Fiend's Fell, but it can be reached most easily, although still with a bit of a hike, from the village of Kirkland in the Eden Valley on the western side of the Pennines. Just below the summit of Cross Fell, this path intersects the Pennine Way, marked by a convenient stone laid into the path. Turn right and follow the Pennine Way over Cross Fell and down its southern flank and you'll arrive at Tees Head.

The landscape of the high fells reminds me strongly of places I've explored in the Arctic. Stone polygons litter the ground, the result of frost heaving. Repeated cycles of freezing and thawing gradually lift stones from the soil and deposit them on the surface in patterns that look far too regular to be natural. If a reminder is needed of the Arctic climate up here, then this is it. The ground is crisscrossed by small streams, and it would be hard to pick out one to call the Tees if not for an old marker stone, looking remarkably like a gravestone, etched with the letters 'B/T' on one side and 'F' on the other – signifying the boundary between the manors of Blencarn and Thanet and the land of a Mr Fleming. The Tees here was also once another boundary, between the old counties of Cumberland and Westmorland. Close to its source, the river is small enough to step across, although not much further downslope, so many tiny streams have coalesced that the Tees is substantially wider.

I began my journey at the other end of the river with a somewhat grandiose comparison of the Tees and the Mississippi and here, at the end of my journey, I'm yet again reminded of similar travels I've made along 'America's Waterway'. I once visited the source of the Mississippi, which, I must say, is a lot easier to reach than that of the Tees. There's a convenient car park close by and a large sign – well... that's America. The Mississippi is born in Lake Itasca in the vast network of bogs, lakes and forests in northern Minnesota, but originally the mighty river began as little more than an indistinct seepage through marshy vegetation at the lake's edge. Not especially fitting for such an iconic river. So, the Army Corps of Engineers, which built the massive control structures at the other end of the river above New Orleans to prevent the river changing course, tackled the smaller job of making the birth of the Mississippi a little more dramatic. The river now begins in a little trickle over a low wall of stones – even though this isn't actually the original source of the Mississippi. It was moved to make the new construction more accessible. Thankfully, there are no similar plans for the birth of the Tees, in what feels like one of the most remote spots in England.

In fact, the view from up here is far superior to the view from the source of the Mississippi. The little gully that marks the path of the young Tees leads off into a vast landscape of open fells. There's almost no sign of civilisation, apart from the radar station perched on the summit of neighbouring Great Dun Fell, part of the air traffic control system. Yet, as we've seen along our eighty-five-mile journey, there's not much truly natural in this – or any other – view along the Tees. The plants of Upper Teesdale may be a direct link to the end of the last glacial period and the start of a more benign period called the Holocene, but this was the last time these landscapes were truly wild. Since then, and throughout the 10,000-year stretch of the Holocene, the influence of humans has gradually increased until, beginning in the middle of the twentieth century, our impact on the world became so prevalent and inescapable that we entered a new geological age – the Anthropocene. And that's not good news for anyone – anywhere on the planet.

From our overview on the lofty flanks of Cross Fell, we can look back down the river along which we've travelled – at its natural world and its social history – and start to draw all these threads together. After all, the value in travelling lies in the journey, not the destination. So, what have we learned from our eighty-five-mile trek across this corner of North-East England? And what can we take from this journey to guide us all on our journeys into the future?

Epilogue

Reflections on Humans
in a Natural World

The Holocene has ended. The Garden of Eden is no more. We have changed the world so much that scientists say we are in a new geological age: the Anthropocene, the age of humans.

DAVID ATTENBOROUGH, Address to the World
Economic Forum, Davos, 2019

That country is the richest which nourishes the greatest number of noble and happy human beings.

JOHN RUSKIN, *Unto this Last*, 1860

A land ethic changes the role of Homo sapiens from conqueror of the land community to plain member and citizen of it. It implies respect for his fellow-members, and also respect for the community as such.

ALDO LEOPOLD, *A Sand County Almanac*, 1949

It's close to six decades since I discovered my love for nature on the banks of the Tees. In that time, in pursuit of making wildlife, environmental and science documentaries, I've travelled the globe to work with scientists, naturalists and conservationists of many disciplines and I've spent time with indigenous and traditional societies in remote rainforests, deserts and mountains, from nomadic Bedouin in the Negev Desert to the forest-dwelling Kayapo and Arara peoples deep in the Amazon. Those experiences have shaped my perspective on our place on this shared planet, and on our responsibilities as citizens of that planet, both to each other and to the natural world. To extend the thoughts of Aldo Leopold that begin this Epilogue, it's become clear to me that we need a new 'nature ethic' as well as a new 'social ethic', and we need them quickly. In fact, as we'll explore shortly, these two aspects of life are sides of the same coin.

We've just travelled along one short and not-very-famous river, yet the stories and ideas that we've uncovered are a microcosm of those playing out on a much larger scale the world over. Such local stories are often more powerful and resonate more strongly than remote global ones and so are perhaps the most immediate way for us to understand the problems we face in the modern world, both socially and ecologically. For that reason, throughout our journey I've painted these stories onto a much larger national and international canvas. In this Epilogue, we'll revisit parts of our journey to weave together the separate strands that we've uncovered into a single narrative. In part, this book has been autobiographical, so what emerges is a very personal perspective, shaped both by my enduring links to the North-East, its people and its wildlife, and by my travels around the globe over five decades.

Among all the natural and social histories that we've encountered, along the whole course of the river and through most of its history, two features have recurred constantly. They are environmental degradation and social inequality. It might seem that these topics belong to two entirely separate discourses, but they're inextricably linked. One study of fifty countries showed that the more unequally a nation's wealth was distributed among its population, the more the biodiversity of that nation was also threatened. Social inequality erodes social capital – the trust of the population and its willingness to work together to demand better environmental health. In addition, societies with greater inequality suffer more crime, drug-use, teenage pregnancy, mental illness and obesity. In the end, inequality also damages democracy itself, since a small, ultra-wealthy elite holds the power to influence politics for its own continued benefit. For example, the world's five largest publicly owned oil and gas companies spend about $200 million per year on lobbying to control, delay or block binding policies to tackle climate change.

In parallel, environmental degradation threatens to pull the rug out from beneath all our feet. Ultimately, we depend on the natural world for our survival, yet we've so damaged the planet that survival is no longer a certainty – at least not without some prompt and massive efforts and a paradigm shift, an entirely novel way of looking at nature, politics and economics that pulls all these apparently separate strands together into one coherent worldview. The twin problems of social inequality and the ecological crisis must be solved in parallel, but before we ask how we move forward, let's remind ourselves how we got here in the first place.

The view from Cross Fell as well as the unique flora of Upper Teesdale are reminders of a time when the ice sheets of the last glaciation had only just retreated, leaving much of Britain covered in tundra. However, as the climate continued to warm, trees arrived and forests covered much of Britain. A period that geologists call the Holocene had begun – and it turned out to be a very special period. We owe the existence of all our societies, right around the planet, to the Holocene because, for 10,000 years, it was an unusually stable and benign period. The global average temperature never varied by more than 1°C and atmospheric carbon dioxide was firmly fixed at 280 parts per million. One reason for this seems to be the way the Earth's orbit around the sun shifts over time.

It's been known for some time that the periodic advance and retreat of ice sheets – the cycle of glaciations and inter-glacials – is caused by the way the Earth wobbles on its axis, along with the way its orbit shifts between a more circular to a more elliptical path.* These oscillations cause tiny changes to the amount of sun that bathes the landmasses of the Northern Hemisphere† each summer and, acting through feedback loops that amplify the changes, these variations are enough to drive the ice back to high latitudes or to cause it to advance again. During the Holocene, the Earth's orbit has been unusually cir-cular, so the variations in the intensity of sunlight bathing the Earth are smaller than when our planet is following a more elliptical path. This last happened around 400,000 years ago and astronomers predict that the benign Holocene could have continued for another 50,000 years – except that we came along.

Not far from Langdon Beck, there's a cave high on the slopes of Langdon Common, on the north side of the Tees. It looks south towards the Tees, and beyond that to the whin-stone cliffs of Cronkley Scar. It's variously called Moking Cave, Langdon Beck Cave, Teesdale Cave or Backhouse Cave, the latter because it was excavated in the 1880s by the Backhouse family of Darlington and York, who, apart from their interests and expertise in plant-breeding, botany, ornithology and bank-ing, were also archaeologists and geologists. Most of their finds are now in the Yorkshire Museum in York and these, along

---

*   These oscillations are known as Milankovitch Cycles, after the Serbian astronomer Milutin Milankovitch.
†   Land warms more quickly than the ocean, and since there's more land in the Northern Hemisphere, this part of the globe plays a bigger role in amplifying the oscillations of the Milankovitch Cycles.

with subsequent explorations, suggest that the cave and the flat grassy area in front were home to families of hunter-gatherers that moved into Teesdale early in the Holocene.

People first came here in the wake of the retreating ice, a time when reindeer and musk oxen roamed the tundra of the Tees Valley. The generations that followed witnessed the slow northward march of trees and these later Teesdale families hunted deer, wild boar and aurochs on the forested hillsides. Lynx also hunted deer in these forests and the bones of these long-vanished cats have turned up in Teesdale Cave. The Mesolithic hunter-gatherers of Teesdale probably also speared the salmon that would have choked the Tees during their spawning migrations.

Millennia later, new families moved into the area around Teesdale Cave, with a new way of life. These people were farmers, and they began clearing patches of trees and enclosing fields with stone walls. Some botanists think that their activities may even have helped the survival of the unique Teesdale plant communities which depend on open ground to thrive. However, agriculture gradually brought sweeping changes to the landscape, not just in Britain, but across the whole planet. Early in the Holocene, agriculture evolved independently in many separate parts of the world – 11,000 years ago in Mesopotamia, 10,000 years ago in China and Central America, 8,000 years ago in India, Africa and the Andes and about 7,000 years ago in the highlands of Papua New Guinea. We can thank the unusual stability of the Holocene for the development of agriculture in all these places and for a climate conducive to building complex civilisations on the back of farming.

The farmers of Britain inherited agricultural traditions developed in the Fertile Crescent in Mesopotamia, and along our journey we've traced an agricultural revolution from these first farmers, with small fields carved out of the forest, to the industrial farmed landscapes of the modern day, which cover 70 per cent of England's land area. Globally, half of all the land on Earth is used for farming, so it's clear that how we manage farmland in the future is critical in solving the ecological crisis that we face. A new approach to farming must be an integral part of our new mindset.

As agriculture and civilisation spread and grew, the distribution of land moved from a more or less equitable division among the population to one skewed in favour of a small elite. Most of our journey has been through land that, following the Norman invasion, once belonged to just two families – de Brus and Balliol. We've also walked through landscapes shaped by successive Inclosure Acts, when remaining common land was taken into private ownership – essentially a huge sell-off of public land into the hands of the wealthy few, which, in the process, created a lot of landless workers who were forced into working their landlord's fields for a pittance or, as the Industrial Revolution kicked off, heading into town to work in factories – still for a pittance. The working class was born in what historian E. P. Thompson (1924–93) called 'a plain enough case of class robbery'.*

If this land grab created much of the social inequality that we're still living with today, the way that the land was farmed also helped create the parallel ecological crisis. Part of this crisis stems from the way we use fertilisers, such as those based on nitrogen, an essential nutrient for plant growth. Soil bacteria can 'fix' atmospheric nitrogen, turning it into ammonia compounds that plants can absorb and use to grow. One family of plants, the legumes, have harnessed these bacteria and grow colonies of them in nodules on their roots. This allows these plants to make their own fertiliser and, when they die and decompose, to fertilise the soil. This is all part of the natural cycle of nitrogen, as this element, bound into different compounds, passes from the atmosphere, through the soil and living

---

* E. P. Thompson. *The Making of the English Working Class*. Random House. 1964.

organisms, and eventually back into the atmosphere, where the cycle begins again.

Farm crops need a lot of nitrogen, and when those crops are harvested and sent to market, this vital nutrient is depleted from the soil. It can, however, be replaced by fertilising the fields. As we saw in the open fields of the Anglo-Saxons, this has been accomplished by rotating crops on each field such that legumes, which return nitrogen to the soil, were grown in one cycle. The dung of domestic animals was also used. In this case, the acreage of crops that could be grown was therefore determined by the numbers of livestock that were reared (which in turn was determined by the acreage of fields of grass and hay to feed them) and this limited the scale and productivity of these farms. It also necessitated a diversity of landscapes across each farm. Importantly, however, much of the nitrogen still cycled locally, as it always had done. Later, guano (the nitrogen-rich droppings of seabirds), imported from around the world, was used, and crop yields increased dramatically. The bird-shit business boomed, and fortunes were made, although, in the process, many seabird colonies around the world suffered catastrophically. In addition, nitrogen shipped in from half a world away began to transform and distort the nitrogen cycle on farmed land. That transformation went into overdrive when Fritz Haber and Carl Bosch invented a way of copying nitrogen-fixing bacteria and making ammonia fertilisers from atmospheric nitrogen.

These game-changing fertilisers were first made a little over a century ago, in 1913, by BASF in Germany. On the Tees at that time, the town of Billingham, close to the industrial centres of Middlesbrough and Stockton, had yet to be swallowed up in the expansion of the heavy industries, but in 1918 the government approved the building of the Government Nitrogen Factory there. Ammonia can also be used in explosives[*] and a

---

[*] The German plant was destroyed in an explosion in 1921.

recent world war prompted the development of the factory, but it was as a manufacturer of ammonia and nitrate fertilisers that Billingham would play its part in transforming farming landscapes and in accelerating a global ecological crisis. In 1926, it became part of the newly created Imperial Chemical Industries (ICI), which later acquired a site at Wilton, further down the Tees between Eston and Redcar, to produce plastics.

Shortly after it opened, Aldous Huxley visited the Billingham site, where its hi-tech tangle of pipes, storage tanks, railway lines and chimneys helped inspire the dystopian vision of the future portrayed in his 1931 novel *Brave New World*. The book was set in 2540, but, even by the twenty-first century, we've already managed to achieve our own version of a brave new world. Indiscriminate and over-use of fertilisers has ushered in the most significant changes in the nitrogen cycle in perhaps 2.5 billion years. We'll come to why this is important shortly.

The modern farming landscape owes as much to subsidies as it does to overuse of chemical fertilisers, especially since the Agriculture Act of 1947, which was designed to maintain high agricultural productivity by guaranteeing minimum prices. The Act was successful in that, in the years following the Second World War, it did increase production, but at the expense of the rapid destruction of a more diverse British countryside. These changes accelerated after 1973, when Britain became part of the European Union. The Common Agricultural Policy (CAP) commanded fully one third of the entire EU budget and was possibly the single most destructive force to hit the British countryside since it was wiped clean by the ice sheets of the last glaciation.

Farmers received payments for every head of livestock, which encouraged overstocking – and the ecological damage which followed. Despite the wild appearance of the high fells where I'm sitting, much of the land here is as degraded as the industrial land around the estuary. It's been so overgrazed and trampled by sheep that very little diversity exists in the short-cropped turf. The CAP also encouraged more meadows to be ploughed up and replaced with heavily fertilised ryegrass-clover leys, as haymaking was replaced by sileage-making, which then allowed even higher stocking rates of cattle and sheep. It paid to bring even the most marginal land into cultivation, even if that meant spraying it with herbicides and insecticides from helicopters. As always, though, the cut of the substantial subsidy pie was not equal for all. The biggest farmers, who were making the biggest profits, also received more of the subsidies. Something like 80 per cent of the CAP budget went to 20 per cent of the farmers.

This all-out and blinkered focus on production resulted in 'wine lakes' and 'butter mountains' that had to be sold off outside the EU at a loss. At the same time, other CAP subsidies were meant to encourage wildlife-friendly farming, often in direct opposition to the payments intended to encourage high production. It's only recently that significant funding was made available to redress the damage done by earlier CAP policies and, once again, it's largely too little, too late. Most of the damage has been done. The EU's conservation directives, such as the Habitats Directive and the Birds Directive, did encourage broad environmental and wildlife protection through the establishment of Special Areas of Conservation and Special Protection Areas, but their measures were often overwhelmed in the face of massive agricultural subsidies. In any case, in Britain, a lot of these sites, many of which are also Sites of Special Scientific Interest (SSSI) listed under an earlier British legal designation, are not fit for purpose. In 2021, nearly one third of SSSIs were discovered to be in 'unfavourable' condition, meaning that their ecological value had been lost or much diminished.

In 1992, under the MacSharry Reforms, the destructive headage payments of the CAP subsidies were changed to ones based on acreage, which at least created less incentive for high stocking levels, although we're still living with the ecological consequences of former policies. In fact, our whole approach to how funding shapes rural landscapes is still a bit of a mess and it buries farmers under yet another surplus mountain – this time, of bureaucracy. It's also bound up with the obsolete economic model that lies at the heart of ecological degradation and social inequality, both locally and on a global scale.

Earlier in our journey, we climbed the steep slag hills of Maze Park near Thornaby, from where we could see the site of the old Head Wrightson ironworks, the place visited by Margaret Thatcher in 1987. From this vantage point, we could also see the results of her deep faith in the economic model known as neoliberalism. The environmental campaigner George Monbiot considers neoliberalism to be the core reason for our current social and ecological problems, and he's not alone. After forty years of neoliberalism, the International Monetary Fund, no less, recently declared that this approach is jeopardising the future of the world economy.* Earlier versions of this philosophy were certainly responsible for driving the Great Acceleration that flipped the planet from the stable Holocene into the uncertain Anthropocene in the 1950s. However, as we've seen along our journey, these ideas have much deeper roots. Thatcher may have been one of its most vigorous exponents, but modern neoliberalism originated in the 1940s with the work of the Austrian economist Friedrich von Hayek. In turn, von Hayek's work evolved from earlier attempts to build an economic theory that could explain and therefore predict economic cycles with the same accuracy as Newton's theory of gravity had done for the cycles of planets around the sun – and therein lay the problem.

The Scottish economist and social philosopher Adam Smith (1723–90) was instrumental in the rise of ideas promoting the importance of economic liberalism. In an attempt to make economics a predictable science, he constructed models in which individual people were assumed to be driven largely

---

* Mark Maslin. *How to Save Our Planet: The Facts*. Penguin. 2021.

by a desire to accumulate material wealth, basing their decisions on the best way to do this on market prices. An entirely unhindered and free market was therefore seen as the most powerful tool for controlling trade by setting the prices of those commodities sought by an insatiable populace. Others built on this, advocating an even greater degree of 'laissez-faire' economics, and created models that further simplified individual human motivations in more extreme ways until we were left with consumers motivated solely by greed and selfishness with perfect knowledge of the marketplace to achieve their goals. Thankfully, most people are a long way from such a stereotype, a fact only too obvious even to the fathers of these economic models. Smith realised that while enlightened self-interest was 'of all virtues that which is most helpful to the individual',* he also recognised that humanity, justice, generosity and public spirit were vital ingredients in the human psyche, qualities which unfortunately made individual behaviour hard to predict and so were excluded from the developing economic theories.

In the following century, John Ruskin (1819–1900), perhaps better known as an art critic, roundly dismissed the evolving economic models of his time as 'mad' – about as meaningful as 'a science of gymnastics which had as its axiom that human beings in fact didn't have skeletons'.† He considered political economy a false science because it failed to include social capital – those aspects of human nature that bind individuals into communities. Ruskin, who also recognised the close relationships between nature and society, prefigured much modern environmental thinking. He argued that the basis of political economy should not be labour and capital, production and consumption, but pure air, water and earth. Almost exactly 150 years later, in 2022, the United Nations General Assembly

---

* Adam Smith. *Theory of Moral Sentiments.* J. Beatty and C. Jackson. 1759.
† John Ruskin. *Unto This Last.* Cornhill Magazine. 1860.

adopted a resolution recognising a universal human right to a clean, healthy and sustainable environment.

Ruskin was a man far ahead of his time, and I like to think that he mulled over these thoughts, much as we're doing now, while staring out over the wild landscapes of the Pennines.

He was a regular visitor to Farnley Hall in the Yorkshire Dales to the south of Teesdale, and clearly loved these landscapes. He described the view from Kirby Lonsdale, on the western edge of the Pennines, as 'one of the loveliest in England and therefore in the world'. This same view was captured in sketches by J. M. W. Turner in 1816 and described by William Wordsworth in his 1810 *Guide to the Lakes*, although it's usually still called 'Ruskin's View', even on local road signs. Ruskin's political views, however, called into question the assumptions that underpinned the unbridled drive for growth and expansion of the Industrial Revolution.

Despite these sharp criticisms of laissez-faire economics, economists ploughed on, building the foundations of neo-liberalism. Key to this philosophy is the unquestioned belief that a successful economy is one that is always growing. That being the case, what was needed was a metric to measure this growth and thereby judge the success of economic policies. In the mid-1930s, the US Congress commissioned the economist Simon Kuznets to come up with a way of measuring America's national income. The result was the concept of gross national product (GNP), which totted up all the income from both residents and overseas US nationals. The concept was later refined to gross domestic product (GDP), which measured just income earned domestically. The idea was enthusiastically adopted by nations around the world, who now focused every effort on making GDP grow as fast as possible, no matter the social or ecological consequences.

Like the portrayal of human consumers in the underlying economic models, GNP and GDP are crude tools, only capturing part of the picture. The enormous value of both social capital and 'ecosystem services' is excluded. Just as Smith recognised the shortcomings of his model of 'economic man', so Kuznets fully understood the limits of his concept. Indeed, by the 1960s he was one of its most outspoken critics. Again, mainstream thinking ignored the concerns of those who really understood the problems, not the least of which is that unlimited growth on a limited planet is impossible. Furthermore, while the free market is a powerful tool for setting prices, not all the costs are built into the model. Some, the so-called 'externalities' such as pollution and environmental degradation, the extraction and depletion of limited resources and the reduction of biodiversity, are excluded. So, the market has no way of balancing these effects with production and demand.

Two decades ago, the environmentalist Jonathon Porritt argued that capitalism as such isn't the enemy.[*] It just needs to incorporate all the costs, including the externalities. More recently, economists have started to look at models that account for many more of these factors and which also have a very different goal from a simplistic target of continued GDP growth. Back at Maze Park, we discovered the work of Oxford economist Kate Raworth and her concept of 'Doughnut Economics'. Unlike the current model, this one has more broad-ranging targets. The bottom line is not continued growth (although that might happen incidentally) but the elimination of social inequality and environmental degradation. This might sound like a pipe dream, and, although Raworth admits that there's a lot that we don't know about making this a reality, she's plotted a course that sets us firmly in the right direction.[†]

What we do know are the limits within which this model must work. All the way along the river, we've seen the problems caused by unchecked exploitation of land and people. Through

[*] Jonathon Porritt. *Capitalism as if the World Mattered*. Earthscan. 2005.
[†] Kate Raworth. *Doughnut Economics: Seven Ways to Think Like a 21st Century Economist*. Chelsea Green Publishing. 2017.

the centuries, these effects became more and more all-pervasive until, during the Great Acceleration of the 1950s, we thrust ourselves from the benign Holocene into a much more uncertain Anthropocene. If we're to thrive or even survive in this future, we need to work within the limits of the resources offered by the planet. A decade and a half ago, Johan Rockström, at the Stockholm Resilience Centre, led a team of twenty-eight scientists to identify these critical planetary boundaries.

They discovered nine boundaries that, if crossed, could act as tipping points by triggering runaway positive-feedback loops, leading to catastrophic changes that would make our lives a lot more difficult, if not impossible.[*] The big three are the climate system, the ozone layer and ocean temperature and acidity. Other boundaries which the best science warns would be foolish to cross include those set by biodiversity and nutrient cycles. Time is running out. We've already crossed three boundaries, including these last two. Biodiversity loss is destabilising ecosystems worldwide and the nitrogen cycle has spun out of control since Fritz Haber and Carl Bosch invented the process for mass-producing fertilisers. Nitrogen pollution (which also results from the vast quantities of manure produced by livestock) has wide-ranging impacts on all ecosystems, from rivers and lakes, where it encourages lethal algal blooms, to woodlands, where it causes soil acidification and the death of trees. In both cases, biodiversity is dramatically reduced. It also affects our own health since it plays a role in smog formation and has even been linked to asthma and some forms of colon cancer. Most of us are now aware of problems arising from our carbon footprint, but those from our nitrogen footprint are just as important.

We ignore these limits at our peril, yet progress towards working within a safe space for all these boundaries is still

[*] Johan Rockström and Owen Gaffney. *Breaking Boundaries: The Science of Our Planet*. Dorling Kindersley. 2021.

limited at best. To stay within the climate safety margins, for example, we need to halve carbon dioxide emissions by 2030 and then halve them again by 2040, and then once more to achieve net-zero carbon emissions by 2050. In this case, however, there is some cause for cautious optimism. The International Energy Agency is monitoring the switch away from fossil fuels to renewable energy sources and, as of 2023, its data suggest that globally the increase in renewables is more or less on track to limit climate change to a rise in average temperature of 1.5°C. That's a boundary set by climate scientists at which, while still causing considerable changes to the planet, we might avoid the very worst effects of rising global temperatures.

Unfortunately, other models suggest we're going to overshoot this target. For example, July, August and September 2023 were all the hottest of those months on record, and September temperatures jumped by an astounding 0.5°C, described as 'absolutely, gobsmackingly bananas' by scientists at Berkeley Earth, a climate data project in California. That just about says it. In the end, September 2023 turned out to be the hottest since records began in 1850 and indeed probably the hottest since the previous interglacial, a time when hippos and rhinos roamed over the area that is now London. The world will inevitably change – there's no going back – but we still have time to halt the worst effects of climate change... just.

In 2020, the world generated $88,000,000,000,000. Economists estimate that we could solve climate change now by spending 1 per cent of world GDP. Or we could wait until 2050, when it would cost over 20 per cent of world GDP. It's not that we can't afford to fix climate change now; rather, we can't afford not to. So, it was depressing to see the former Tory government issuing new oil and gas exploration licences and, as of 2024, the incoming Labour government not rescinding these (although it has ruled out issuing further licences). We also need to rebuild lost biodiversity. In 2018, the UK government announced ambitious plans to restore nature across the country and to protect 30 per cent of the land by 2030. Unfortunately, the Office of Environmental Protection, which examined twenty-three of the government's targets, discovered that it was on track for none of them and clearly well off-track on fourteen of them. The government also ignored advice from its own expert body, Natural England, on how best to achieve those targets. However, one ambitious target, to restore hedgerows grubbed out as farming intensified, went well beyond conservationists' recommendations and had them rejoicing – until it turned out to be a typo in the report. Oh well.

When it comes to restoring nature, we've already seen that farmers have a critical role to play. Since those first farmers turned up at Teesdale Cave, nature has adapted to farming practices. Landscapes managed in traditional ways – coppiced woodlands or meadows grown for hay, for example – were often naturally diverse, as we discovered in Hannah's Meadow in Baldersdale. So, there's no reason why we can't recreate some of this past interplay between people and nature in a countryside of the future. As foolish a decision as Brexit is starting to look overall, it does at least mean that we in Britain can adopt a new agricultural paradigm, free from the destructive elements of the Common Agricultural Policy. But, yet again, the government at the time made encouraging noises and then failed to follow through. In 2020, it promised a new 'agricultural revolution'. By 2023, only a small fraction of money allocated to new subsidy schemes, including those aimed at improving the natural environment, had been distributed to farmers, many of whom are close to going out of business.

However, it's not all down to agricultural policy, with its subsidies handed down from on high. We all need to be part of this new paradigm for the future. As one example, meat farming is energy-intensive, takes up vast amounts of land and produces mountains of polluting manure. It takes one hundred times the land to produce the same weight of red meat as it does cereals and, in a time when food security is a critical issue, half the wheat that we grow is fed to cattle. In addition, switching from cow's milk to oat milk reduces greenhouses gases by about 40 per cent. While governments are slow to react, growing numbers of concerned people may force change. In the US, for example, oat milk sales grew 686 per cent in 2019, while over the last two decades in the UK, meat consumption per person has fallen by 6 per cent. Of course, any such social changes need governments to respond more quickly and more positively than is usual. Dairy and beef farmers need to be fully compensated for loss of income, and new subsidies should be targeted to encourage nature on any land taken out of production, as farmers move from intensive to extensive farming – as they become stewards of a much richer and more diverse landscape.

There's a fear among hill farmers in particular, whose living is often marginal in these bleak uplands, that such a paradigm shift will spell an end of their way of life. In some ways it will, since business as usual – for any of us – isn't a realistic possibility in a viable future. As I gaze over Cross Fell, the landscapes around me look as wild as if the ice sheets had only just retreated. But, over the last few decades, 65 per cent of upland wildlife (at least, that for which we have enough data to analyse) has declined, with 35 per cent declining severely.* To redress this we need the knowledge and skills in land management of local farmers, even if that means they must forge a new relationship with the countryside.

One element in this new relationship is 'rewilding' – a term heard more and more frequently from environmentalists, and it can be a powerful tool. In its purest form it lets nature take the driving seat by replacing long-lost species and letting them get on with it. Some species, like beavers and wild boar, became extinct in Britain but still exist elsewhere and are being reintroduced to take up the ecological role they once had here.

---

* The Wildlife Trusts' *State of Nature 2013* is a report produced in collaboration by twenty-five conservation and wildlife monitoring organisations.

Others, like wild horses (or tarpan) and aurochs are globally extinct and therefore must be replaced by ecological analogues. Hardy breeds of ponies or cattle can be used to recreate the grazing patterns of our original big grazers and, in the case of cattle roaming on extensive farms with a low stocking density, produce high-quality, free-range beef, which nevertheless could provide enough meat to a public with the wisdom to drastically reduce the amount of meat in our diets.

This form of reduced or non-management can bring swift and beneficial changes. The most famous example is a large farm at Knepp in Sussex, where these practices have filled woods with nightingales and fields with turtle doves, along with many other threatened and vanishing species of the British countryside. Isabella Tree and Charlie Burrell, the people behind this dramatic example, prefer to call the process 'wilding', which is a term I also prefer. Rewilding suggests recreating something that once existed, which, in modern Britain at least, is an impossible dream. Rewilding also suggests a complete return to nature, which, although perhaps possible in a few of the most remote parts of our planet, is also a misconception. We humans, as long as we exist, will remain an integral and influential part of life on Earth. The future for nature is inevitably one shared with humanity.

Another cause for optimism is the enthusiasm with which this approach is being adopted. In Scotland, the Northwoods Rewilding Network is specifically aimed at smaller land holdings, between 100 and 1,000 acres, and already has 60 partners across Scotland, many of them small farms. The project aims to pioneer ways to make these new management models effective in restoring ecology and at the same time commercially viable for the landowners and farmers. The project's partners vary from small farms to larger estates and woodlands, but the land management can be tailored to meet the needs of each individual location.

There are degrees of wilding, so it's not, as some people

fear, a recipe for depopulating the countryside. The RSPB runs two hill farms, Naddle and Swindale, at Haweswater in the Lake District, which have undergone a form of partial wilding. The former local MP, Rory Stewart, voiced the common opinion that rewilding 'leaves no place for humans in the landscape', but that's far too simplistic a vision. The RSPB has drastically reduced the densities of sheep and cows, planted a lot of trees and re-wiggled a stream that had been dredged straight through the landscape. But it's still run as a farm, and the RSPB aims to produce high-quality meat, while at the same time creating space for nature and a place for people to immerse themselves in a revitalised natural world. As a bonus, and contrary to popular opinion, the number of people working on these farms has increased. It's also argued that to provide food security, returning farmland to nature is a foolish idea. Surely, we need farms to feed a growing population? Yet we discovered earlier in our journey that globally we already produce enough food to feed 11 billion people in a world approaching just 9 billion. The problem isn't food production, it's food waste. In America alone, wasted food amounts to $165 billion, four times the amount of food that Africa imports each year.

It's not just hill farms that can benefit from a significant shift in management and attitude. Since 2000, the RSPB has also been running Hope Farm, in the very different landscape of Cambridgeshire. Here, it has pioneered techniques to encourage wildlife while at the same time maintaining a profitable lowland farm business. About 15 per cent of the land was taken out of production and in ten years, breeding bird populations trebled. Birds, bees and butterflies, among many other animals and plants, now prosper and the farm still makes average profits for a farm of its size. Nature isn't as abundant and spectacular as at Knepp, where wilding has been more extreme (even European bison now graze its forests and scrub and it's a key part of a project to return white storks to Britain), but we'll need both approaches for a viable future. Conservation is sometimes seen as a choice between land-sharing or land-sparing, but each is appropriate in different landscapes and, more importantly, there are degrees of sharing. In more marginal farming areas, greater emphasis should be placed on shifting the balance in favour of nature, while food production can be prioritised in more productive areas, where intensive farming techniques, with appropriate controls on pollution of all forms, can maximise yields from the land.

Unfortunately, returning any farmed land to nature, even partially, faces a lot of opposition, not just here in Britain but in places where there's plenty of space to experiment with new methods of land management. By the end of the nineteenth century, bison across the American West had been reduced from herds that numbered in the tens of millions to just a handful of individuals. Their place on the Great Plains has now been taken by vast numbers of cattle. For a film documenting this tragic story and the subsequent recovery of an American icon, I talked with rewilders, historians, conservationists and cattle ranchers across the West, from Texas to Montana. An obvious step suggested by conservationists is to replace some of the cattle herds on the Great Plains (which, like the sheep of the Pennines, have inflicted severe damage on the grasslands) with bison. These indigenous grazers are much kinder to the native grasses and other plants and encourage a more vibrant web of life on the plains, from prairie dogs to pronghorn antelopes. However, among a great many ranchers, along with their state governments, there is unbending resistance to such ideas. It's as if we're admitting that nature has won, that all the hard graft of the pioneers in subduing nature, in bending it to our will, was somehow a failure.

Although there are plenty of individual farmers in the UK who are concerned about environmental issues and who are taking active measures to improve the ecology of their lands, farming as a whole here, as represented by the influential

National Farmers' Union, is as ultra-conservative as those Western ranchers. Agricultural interests, through the NFU, have persuaded the government to water down the effectiveness of environmental grant schemes, demanded large-scale culling of badgers against the prevailing scientific advice and sought to overturn bans on some of the most toxic pesticides.

And here in Britain we're all as fearful of giving nature a bigger role in shaping our ecology as those American ranchers. To create more balanced ecosystems, we need to bring back predators, yet there was an extraordinary amount of hand-wringing over reintroducing even tree-gnawing beavers into Britain. And this despite the fact that the presence of beavers has had almost universally positive effects on biodiversity and flood control. So, it's hardly surprising that plans to reintroduce carnivores have made little progress. Like beavers, lynx could also enhance biodiversity by reducing deer populations that have grown unnaturally large in the absence of their predators. And since these elusive cats would disappear into the landscape, we'd hardly notice their presence, which makes them the most likely candidate for predator reintroduction when we finally pluck up the courage. I can't begin to imagine the fuss over any plans to bring back wolves or brown bears. We Britons have simply got used to living in a land largely stripped of nature. Yet there are now more wolves and brown bears in Europe than in North America, and conflicts with people are surprisingly few. A change in attitude to big nature is another element of the new worldview we must cultivate.

If we are ever to establish this new paradigm, then at its heart must be a much deeper respect for nature. To repeat the words of American environmentalist Aldo Leopold, 'A land ethic changes the role of *Homo sapiens* from conqueror of the land community to plain member and citizen of it.' For too long, nature has been seen simply as a resource to be exploited or a convenient dump for waste and pollution. Used wisely, nature *is* a renewable resource and *can* provide ecosystem services, such as waste removal, but it's even more important to us than that.

Ahead of his time in so many ways, in criticising the economic policies of his day in 1860, John Ruskin wrote, 'There is no wealth but life.' He was right on so many levels. Nature provides the raw materials for our economy, as we've seen so often in our journey along the Tees, whether the iron ore that gave birth to Middlesbrough, the lead ore extracted from beneath the fells of Teesdale, or the coal from Durham and Yorkshire to fuel these industries. The natural world also supported fishermen from Teesmouth to Yarm. Raw materials for short-lived economic booms, such as the craze for feathers in the late nineteenth century, also derive from nature, at the expense of all but wiping out a great many bird species. But, more than all of these, nature is also vital to our wellbeing. Any new paradigm must include an economic model which works to distribute wealth more fairly and to restore social capital and the power of communities. But part of that wealth is a vibrant natural world – there is, after all, no other wealth but life. Here again I've found encouraging signs of the emergence of new perspectives.

John Ruskin, in a further critique of nineteenth-century capitalism and industrialism, expressed the belief: 'That country is the richest which nourishes the greatest numbers of noble and happy human beings.' Long before Ruskin, that's a truth that's been known for at least two and a half millennia. Buddhism was founded on the core belief that fundamentally we're all seeking an end to suffering and a way to achieve true and lasting happiness. And it's nature and community, not money or power, that generate such wellbeing. I've lost count of the number of reputable studies that I've read that back up this long-held view and new ones continue to emerge all the time. They show that, as long as basic needs are met, financial wealth does not equate to happiness. If you're miserable and then win the lottery, the chances are that after a few months of superficial happiness, you'll be miserable again. So, here is another serious shortcoming of using GDP as a measure of how well a nation is really doing. Financial wealth, even if it does ever trickle down from the wealthy few, isn't a true cause of happiness. One country has broken through this limited perspective and established a new metric. The Kingdom of Bhutan now uses gross national happiness (GNH) as an indicator of progress. Perhaps Bhutan is uniquely placed, since its state religion is Buddhism, but we could all learn from this tiny Himalayan country.

However, Bhutan is no longer alone in recognising the failings inherent in judging a nation's progress solely through its GDP. On another imposing mountain range, I've spent time among both the Quechua and Aymara communities in the South American Andes. Our visit to one village, high on the Altiplano, was timed to coincide with a wedding and we duly arrived with perhaps the most unusual wedding gift that I've ever brought to such a celebration – a llama. The festivities lasted several days and involved the whole village. The good people of Vilacayma certainly know how to party. They soon left me behind, although the fact that the village lies at over 15,000 feet in very thin air may have played a part. Bolivia is a desperately poor nation, yet in all the villages that we stayed in, I was left with an enduring sense that each close-knit community had an underlying *joie de vivre* as they eked out a living in the dramatic Andean landscapes by herding llamas across grassy plains and farming potatoes in fields around the village – scenes unchanged since the days of the Incas.

The Bolivians have a phrase for this, *buen vivir*, living well, and recently the government has incorporated this idea as an ethical principle within the country's constitution. In common with many indigenous societies, these people also possess a deep reverence and respect for the natural world that supports them. Along the Andes, nature is personified as Pachamama, and in 2008, Ecuador's constitution recognised that Pachamama 'has the right to exist, persist, maintain and regenerate its vital cycles'. In 2012, Bolivia adopted the Law of the Rights of Mother Earth, granting Pachamama the same rights as a human. Far from being unique, this holistic and balanced approach to life was common among many cultures around the world, who recognised that true prosperity comes from being part of a vibrant community and a flourishing natural world. Even Western countries are now beginning to appreciate this ancient wisdom.

In 2017, New Zealand's government passed a law granting the Whanganui River the same legal rights as a person. Subsequently, several other places in New Zealand have gained similar legal protection. Before you think that the government of New Zealand has gone raving mad, consider that many companies already have such legal rights – so why not important features in the natural world? In 2019, Jacinda Ardern, New Zealand's prime minister at the time, published the country's first 'wellbeing budget', one that looks beyond GDP as a metric for future development. Ardern also worked with Nicola Sturgeon, then Scotland's first minister, and Katrin Jakobsdóttir, Iceland's prime minister, to form a loose alliance called Wellbeing Economy Governments (WEGo). There's also the United Nations' Human Development Index, the Happy Planet Index, the Inclusive Wealth Index and the Social Progress Index, all set up to counter strict adherence to GDP as a measure of progress. So, the good news is that changes are happening even in developed Western economies. The bad news is that from politicians on all sides, we still only hear about the drive for simple economic growth.

We all need a thriving natural world for our current wellbeing as well as our future health and happiness. Therefore, central to our emerging new worldview is the idea that we're just one species in a community of living beings. This isn't a new idea. It's one that we in the Western world have lost as we have become ever more distanced from nature, but it's common knowledge among the American Indian tribes that I've visited

over the years. Among the Lakota Nation on America's northern plains there's an expression that sums this up – *Mitákuye
Oyás'in*. Translated from the Lakota language, it means 'all
are related'. According to the American scholar Joseph Epes
Brown, who spent much time with the Lakota in the 1940s,
it's a reflection of an underlying connection and a oneness with
the world around them. This isn't just some outdated or naive
ideology. It's well known that we share 98 per cent of our genes
with chimpanzees, but we also share 80 per cent of our genome
with the cows on our farms and 60 per cent with the insects
that we are so effectively banishing from our countryside. We
even share 50 per cent of our genome with the crops growing
in those farm fields. The Lakota are absolutely right – all *are*
related.

We come to that understanding in different ways. As I wrote
before we even began our journey, I'm not entirely sure how
a lad from a council estate in Middlesbrough ended up with
a life-long connection to nature, although, looking back over
space and time from my lofty seat on Cross Fell, I can see a few
ways that it might have happened.

As far as I was aware, no one in my family had the slightest
interest in animals or plants until, one day shortly before his
death, I was quizzing a paternal uncle about his early memories of family life in Middlesbrough in the 1920s and 1930s. I
never knew my paternal grandfather and neither did my dad.
He died of tuberculosis when my dad was only a few years
old. However, my uncle was older and, as a child, had a few
clear memories of a man who in photographs bears a striking
resemblance to my father and, as it turns out, to me. Recently,
my three-year-old great niece mistook a picture of him for me.

My uncle told me a story that no one else in the family had
heard before. My grandfather was a labourer in Middlesbrough's
ironworks and lived in a small house on Brompton Street, a
row of tiny houses in the centre of town. Every evening, after
tea, he carried a set of stepladders into the centre of the cobbled

street, climbed to the top and whistled. In this way, he soon marshalled all the local stray dogs and ushered them into his house. He fed them all and gave them a bed for the night before turning them loose early the next morning as he headed to work. This doesn't strike me as in any way eccentric – rather, it was the action of someone who, despite his own poverty, recognised the needs of other creatures, and his own wellbeing was no doubt enhanced by these encounters. I don't know if there's a genetic disposition for empathy with other animals, but I wish I'd been able to get to know him. And I'm proud to have been mistaken for him.

When I was just a few years old, a time from which I have only vague memories, I'm told I spent a lot of time looking for bugs and worms in the garden and asking awkward questions about what they were and what they were doing – so much so that one of my long-suffering uncles only ever called me 'the professor'. Since I can't remember much about this time, I have no idea what sparked this interest. However, I have one important later childhood memory that I still recall clearly. It concerns a trip to Flamingo Park Zoo (now Flamingoland), just to the south of the North York Moors. I was only about eight at the time and we had no car, so the journey involved several bus journeys, making it a bit of an adventure. I was already fascinated by all animals, so my excitement at seeing so many, from lions down to tiny lizards, was boundless, but one encounter stands out.

At the time, the zoo featured a dolphin show, something that now I can't condemn vehemently enough. But for a naive kid it was spellbinding. At the end of the show, the crowds disappeared to find the next bit of entertainment, but I stayed behind. The dolphins were in a pool with completely open access so, since no one else was around, I sat on the edge of the pool. After a few minutes, one of the dolphins came over and stuck its head out of the water just inches from mine. We sat together quietly like this for several minutes and, as I looked into its eye, I had the most extraordinary and powerful sense of a fully sentient mind scrutinising me – a strong connection between two very different species. I still recall that feeling with absolute clarity. It changed the way I see all the animals with whom we share the planet.

Six decades later, I was wandering over the grassy plains of the Fort Belknap Indian Reservation in northern Montana in the company of Mark Azure, an Assiniboine who is part of an inter-tribal project to restore buffalo across Indian Reservations in the plains states, where state legislatures, with their reluctance to allow nature a place in our future, hold no sway. We walked up to a small herd grazing quietly and Mark explained the importance of the project, not just for conservation but for the wellbeing of his and all the plains tribes. 'These are not just animals', he said. 'They're our brothers and sisters.'

Acknowledgements

My thanks to Ken Smith, not only for nurturing a budding naturalist in my youth, but also for commenting on the manuscript. His knowledge of local natural history is legendary and I'm grateful for his helpful thoughts and suggestions. Likewise, I'm grateful to my long-time friend and film-making colleague Paul Reddish, whose knowledge of fish and fishing is encyclopaedic. I thank him for keeping the fishy tales in the preceding pages on the straight and narrow, especially when it comes to fly fishing and to the complex genetics and biology of salmonid fish. Thanks also to Paul Dunn, of Croft-on-Tees, for taking the time to show me around St Peter's Church in Croft and for generously sharing his knowledge of local history, especially on its relevance to the well-known 'Alice' stories of Lewis Carroll.

Many thanks to Pat Garbutt, wife of singer-songwriter Vin Garbutt, who sadly passed away in 2017, for allowing me to use a few lines from his songs, many of which encapsulate the ups and downs of life and nature in the Tees Valley in powerful and emotive lyrics. Vin was an extraordinary performer with a unique voice. I saw him many times in folk clubs around the North-East and his songs still resonate strongly with me. Indeed, they were a frequent accompaniment and inspiration during the writing of this book.

Thanks to Richard Milbank, my editor at Head of Zeus, for his thorough yet sensitive editing. This is the third of my books that Richard has edited and, in each case, the process has been a real pleasure as we've made significant improvements to my original drafts. Thanks also to Kate Hordern, my literary agent at KHLA Ltd., both for helpful discussions over my original back-of-an-envelope ideas for this book and for helping to shape these into a coherent outline. I'm also grateful for Kate's insightful comments on the first draft. Thanks also to the design team, who have, in all three of my projects for Head of Zeus, crafted beautiful books.

I'm enormously grateful to my two sisters, Christine Nicholls and Carole Johnston, life-long denizens of Middlesbrough, for their unending hospitality and congenial company on my frequent visits to the North-East over the years. I also owe an enduring debt of gratitude to my parents, Joan (1922–2008) and Ron (1924–2021), for allowing me to fill our small house in Middlesbrough with an assortment of natural history specimens, both alive and dead, and for their unquestioning support of my growing passion for the natural world, even if they never fully understood where it came from or where it would lead.

Finally, thanks to my wife, Vicky Coules, for her unfailing support throughout this whole lengthy project, especially when the enthusiasm of beginning and the satisfaction of finishing both seemed a long way away. She also read the early drafts and helped shape the journey we take through the book by curbing my frequent and lengthy sidetracks into areas of personal fascination which perhaps my readers wouldn't share. But more than that, Vicky is a talented artist who has accompanied me on many trips along the Tees. I'm enormously grateful for the line drawings she produced on those visits to augment my photographs in illustrating the text.

Bibliography

Local History and Local Natural History

Allan, Dave. *The Transporter: 100 Years of the Transporter Bridge.* Middlesbrough Council. 2011.

Amor, Anne Clark. *Lewis Carroll: Child of the North.* The Lewis Carroll Society. 1995.

Blick, Martin A. *Birds of Cleveland.* Tees Valley Wildlife Trust. 2009.

Bradshaw, Margaret E. *Teesdale's Special Flora: Places, Plants and People.* Princeton University Press. 2023.

Carson, Robert. *A Short History of Middlesbrough.* Middlesbrough Borough Council. 1977.

Chohan, Araf. *Middlesbrough St. Hilda's.* Destinworld Publishing Ltd. 2015.

Chohan, Araf. *Middlesbrough: A Century of Change.* Destinworld Publishing Ltd. 2019.

Chrystal, Paul. *The Romans in the North of England.* Destinworld Publishing. 2019.

Chrystal, Paul and Laundon, Stan. *Secret Middlesbrough.* Amberley Publishing. 2015.

Clapham, Arthur Roy, ed. *Upper Teesdale: The Area and Its Natural History.* Collins. 1978.

Collins, Martin and Dillon, Paddy. *The Teesdale Way: From Dufton to the North Sea with 10 Circular Day Walks (British Long-Distance Trails).* Cicerone. 2005.

Forbes, Ian, ed. *Eyewitness Accounts of Lead Mining in the Pennines.* Friends of Killhope. 2011.

Garland, Richard. *A Tour of Teesdale Including Rokeby and Its Environs.* 2nd ed. Thomas Wilson and Sons. 1813.

Gater, Steve. *The Natural History of Upper Teesdale.* Mosaic (Teesdale) Ltd. 2018.

Gilbertson, Peter. *Two Days in Teesdale: Dickens' Visit and Its Legacy.* Mosaic (Teesdale) Ltd. 2012.

Hannah Hauxwell and Barry Cockcroft. *Seasons of My Life: Tales from a Solitary Daleswoman.* Century Hutchinson Ltd. 1989.

Hough, Richard. *Captain James Cook.* W.W. Norton & Company. 1995.

INCA. *Wild Teesside: Thirty Years of Industry and Nature.* 2020.

Joynt, Graeme, Parker, Ted and Fairbrother, Vic, ed. *The Breeding Birds of Cleveland.* Teesmouth Bird Club. 2008.

Kirkpatrick, Robert J. *Charles Dickens, Nicholas Nickleby and the Yorkshire Schools: Fact vs. Fiction.* Mosaic (Teesdale) Ltd. 2017.

Le Guillou, Michael. *A History of the River Tees, 1000–1975*. Cleveland County Libraries. 1978.

Moorsom, Norman. *Middlesbrough's Albert Park: History, Heritage and Restoration*. Wharncliffe Books. 2002.

Morton, Margaret and Bradshaw, Margaret, ed. *A Guide to the Natural History of the Tees Bank Wood*.

Pickett, Elizabeth. *Reading the Rocks: Exploring the Geology and Landscape of the North Pennines*. North Pennines AONB Partnership. 2011.

Proud, Keith J. *Charles Dickens in Teesdale: The Story of Nicholas Nickleby and the Yorkshire Schools*. Discovery Guides. 1983.

Rudd, Michael. *The Discovery of Teesdale*. Phillimore & Co. Ltd. 2007.

Shepherd, Cliff. *The Story of Saltholme*. Teesside Environment Trust. 2018.

Simpson, David. *Steel River: Two Thousand Year Journey Along the River Tees*. The Northern Echo. 1996.

Stables, Andrew Graham. *Secret Barnard Castle & Teesdale*. Amberley Publishing. 2018.

Stables, Andrew Graham and Marshall, Gary David. *A–Z of Barnard Castle and Teesdale*. Amberley Publishing. 2020.

Valentine, D. H. *The Natural History of Upper Teesdale*. Northumberland and Durham Naturalists Trust Ltd. 1965.

Warwick, Tosh and Parker, Jenny. *River Tees: From Source to Sea*. Amberley Publishing. 2016.

Watson, Richard. *Poems and Songs of Teesdale*. William Dresser & Sons. 1930.

Wilkinson, Colin. *The Industrial Revolution of the Tees Valley*. Amberley Publishing. 2018.

Woodhouse, Robert. *The River Tees: A North Country River*. Terence Dalton Ltd. 1991.

Woodhouse, Robert. *Stockton Past*. Phillimore. 1994.

Woodhouse, Robert. *Tees Valley Curiosities*. The History Press. 2009.

General Natural History and Environmental Science

Avery, Mark. *Reflections: What Wildlife Needs and How to Provide It*. Pelagic Publishing. 2023.

Barnett, Ross. *The Missing Lynx: The Past and Future of Britain's Lost Mammals*. Bloomsbury Publishing. 2019.

Butfield, Colin and Hughes, Jonnie. *Earthshot: How to Save Our Planet*. John Murray. 2021.

Cocker, Mark. *Our Place: Can We Save Britain's Wildlife Before It Is too Late?* Random House. 2018.

Leopold, Aldo. *A Sand County Almanac and Sketches Here and There*. Oxford University Press. 1949.

Macdonald, Benedict. *Rebirding: Rewilding Britain and Its Birds*. Pelagic Publishing. 2019.

Macdonald, Benedict. *Cornerstones: Wild Forces That Can Change Our World*. Bloomsbury Publishing. 2022.

Maslin, Mark A. *How to Save Our Planet: The Facts*. Penguin. 2021.

Monbiot, George. *Feral: Rewilding the Land, the Sea, and Human*

Life. University of Chicago Press. 2014.

Pearce, Fred. *The New Wild: Why Invasive Species Will Be Nature's Salvation*. Beacon Press. 2016.

Porritt, Jonathon. *Capitalism as if the World Mattered*. Earthscan. 2005.

Rackham, Oliver. *The History of the Countryside*. Hachette. 2020.

Raworth, Kate. *Doughnut Economics: Seven Ways to Think Like a 21st Century Economist*. Chelsea Green Publishing. 2017.

Raye, Lee. *The Atlas of Early Modern Wildlife: Britain and Ireland between the Middle Ages and the Industrial Revolution*. Pelagic Publishing. 2023.

Rockström, Johan and Gaffney, Owen. *Breaking Boundaries: The Science of Our Planet*. Dorling Kindersley. 2021.

Schofield, Lee. *Wild Fell: Fighting for Nature on a Lake District Hill Farm*. Random House. 2022.

Sharpe, Simon. *Five Times Faster: Rethinking the Science, Economics, and Diplomacy of Climate Change*. Cambridge University Press. 2023.

Shrubsole, Guy. *Who Owns England?: How We Lost Our Green and Pleasant Land, and How to Take It Back*. William Collins. 2019.

Smith, Malcolm. *Ploughing a New Furrow: A Blueprint for Wildlife-Friendly Farming*. Whittles Publishing. 2018.

Tree, Isabella. *Wilding: The Return of Nature to a British Farm*. Pan Macmillan. 2018.

Vera, Franciscus Wilhelmus Maria, ed. *Grazing Ecology and Forest History*. CABI Publishing. 2000.

Image List

Index